Systems Project Management

Systems Project Management

EDITED BY

Don Yeates

Pitman

PITMAN PUBLISHING LTD
128 Long Acre, London WC2E 9AN

A Longman Group Company

© Don Yeates 1986

First published in Great Britain 1986

British Library Cataloguing in Publication Data

Systems project management.
1. System analysis
I. Yeates, Don
658.4'032   HF5548.2

ISBN 0-273-02388-8

Computer typeset by SB Datagraphics, Colchester, England.
Printed at The Bath Press, Avon

Contents

Preface

This book is intended for everyone who would like to see systems projects implemented on time, within budget and to quality. While this probably means every user of a computer-based system and every member of a data-processing department or systems house, it is unreasonable to expect so wide a readership. Specifically, then, the book will be useful to:

- Practising systems analysts who find themselves responsible for managing systems projects. Newcomers to this activity will find much that is new; I hope also that some of the older hands will rediscover a few forgotten methods or find some new ideas that will help them to tackle the job with renewed vigour.

- Students of data processing, information systems and project management. Not everything can be learnt from books. Oscar Wilde said that 'experience is the name everyone gives to their mistakes', and throughout the book I have included advice based on experience learnt the hard way.

- Part-time developers. Many people are drawn into the development of application systems and have no need to understand all the duties of the project leader. Selected reading can, however, help them to understand how their activities fit into the whole scheme and will, I hope, lead to better project management.

The book begins with an overview of the role of the project manager and the place that systems projects occupy in the life of an enterprise. Chapter 2 reviews the stages of project development from feasibility to post-implementation support and describes the deliverables to be produced at each stage.

Chapters 3, 4 and 5 address project planning, estimating and project control – often thought to be the main tasks of the project leader. Chapter 6 supports these activities with a detailed review of project administration.

Chapters 7, 8 and 9 tackle the project leader's responsibilities as a leader of people. Choosing the team and the use of appropriate selection methods is followed by team motivation and performance improvement.

Chapter 10 deals with managing quality. A good-quality system is not something that happens by accident. The use of structured methods as an aid to project management is also covered here.

Finally, in chapter 11 the human and sociological implications of technological change are considered in the wider context of organizational change.

There is, inevitably, much that has had to be left out of this book, but I hope that readers will find that what has been included helps them to understand a little more about systems project management.

Don Yeates
Ealing 1986

Acknowledgements

One of the most endearing characteristics of the computer business is the willingness with which people share their experiences — both good and bad. This is particularly true, I think, for people concerned with training, whether they are full time trainers or line managers who regularly participate in company training programmes. This book brings together the practical experiences of many friends and colleagues who have been prepared to share their experiences with me — and consequently with you, the reader. Without them this book would not have been prepared. I am most grateful to them all and would like to thank:

- Dr. Frances Clark, Head of Behavioural Studies (Advanced Information Technology Group), of Coopers and Lybrand Associates, for managing the implementation.

- Martin Hanslip, Contracts Resourcing Manager for the Commercial and Financial Division of Software Sciences — a THORN EMI Information Technology Company, for his work on project management methods, techniques and tools.

- Alan Staines, Education Centre Manager and Paul Wyman, General Manager, Commercial Division, both of Datasolve — a THORN EMI Information Technology Company. Alan for quality management and Paul for his work on the people issues.

Needless to say, their contributions are personal and not corporate, and should not be taken as necessarily representing their employers' views.

Finally, I must thank two long-standing academic friends, Professor Frank Land of the London Business School and Dave Hatter of North East London Polytechnic, for their creative ideas, helpful comments and breadth of vision regarding the problems of project management.

In the end, however, any mistakes are all mine. The word processing of it all was masterminded by Paulette Sharkey.

Don Yeates

1 The analyst as project manager

1.1 Introduction

The project manager's job is the single most difficult, and most rewarding, job in the computer business. This is true whether you work for a computer user, a service company or a manufacturer of computers. Without a project manager, bad systems are implemented late and over budget. Sometimes there is no doubt that this happens with a project manager, but the single most over-riding characteristic of a project manager is that he – or she – achieves goals, reaches targets and delivers on time. The buck stops with the project manager. There is consequently considerable pressure on the project manager from a variety of sources. Expected to manage technology, people and the change process, the project manager can face the following potential difficulties even before the development process for a new system has begun:

- dissatisfaction of users with previously developed computer-based systems;
- project management methods and organization structures inappropriate for new systems development;
- inadequately skilled systems development staff;
- poorly developed systems planning mechanisms.

The project manager* thus finds himself buffeted from all sides by a variety of factors outside his control. Project managers are, however, expected to absorb this pressure and create an environment within which their project team will deliver a successful new application system to time, quality and budget. This chapter is not concerned with

* Throughout this book we refer to the project manager as 'he' because there is no convenient neuter pronoun.

the detail of how this is done: these matters are covered in subsequent chapters. At the beginning of the book, we are concerned with the role of the project manager and the framework or environment within which systems projects are developed. It is an opportunity to step outside the day-to-day activities of applications system development and consider some of the more strategic issues. In particular, we can examine:

- the challenges posed by information technology;
- the need to plan for the application of information technology;
- the role of the project manager;
- how systems change takes place.

Some of this will seem far removed from actually designing a new system, devising an acceptance test plan, motivating a reluctant programmer or influencing a truculent user. This is because it is too late then to consider the strategy when you are in the thick of systems implementation. If you survive, however, you might like to think about it next time!

1.2 Systems projects in context

The role of technology in society is to improve the quality of human life, whether in monitoring the condition of hospital patients, controlling mass transit transport systems, checking the quality control of consumer products or evaluating the earth's resources via satellite. Human development depends increasingly on the availability and the application of technology. What contribution can the application of information technology make to improving the quality of human life? Detailed predictions are clearly very difficult to make but there are three broad areas of technological development which will have a significant impact on our working and personal lives.

Communications technology now makes it possible for cheap and fast electronic mail to cross many of the barriers created by the geographical separation of peoples. This facility, together with text and teleconferencing networks, can allow individuals and organizations rapid communication with each other from any location in the world. These kind of developments clearly have implications for any community or individual wishing to share or exchange information or to collaborate with others, whether their interests are commercial, scientific, political, educational or personal. Information technology can also help us to make better decisions and to solve problems more efficiently. Good decisions should take into account all the relevant information about costs and benefits, should identify all the possible options and their implications, and establish the most appropriate course of action. Typically, however, most decisions rely on human memory or paper-based information systems and are, in consequence, vulnerable to the weaknesses of these often slow and fallible sources of data. Using information technology, we have the ability to store and present all the necessary information in a comprehensible form and to provide sophisticated modelling tools to enable decisions to be based on more comprehensive data and with prior knowledge of the implications of any projected changes which will result from the decisions we make. While we cannot yet see computer-based decision-making systems replacing the human decision-making process, the area of decision support systems is one of the most rapidly growing areas of computer application and focuses exclusively on improving a manager's ability to make better decisions. Of particular significance in the area of improving decision-making will be future developments in intelligent knowledge-based systems (IKBS). These expert systems assemble all the available knowledge in the particular field in a form that can be used for problem-solving by specialists in that field, using their language as opposed to the language of computing. At the moment, they are particularly used in research, in development and design projects in aeronautics and civil engineering, and also for problem diagnosis, especially in medicine. In effect, they create new knowledge by highlighting previously unknown interactions between the variables of a problem. IKBS systems are potentially of great value as they provide support for the limited capabilities of human memory and data manipulation. In the United Kingdom, the government-sponsored Alvey Directorate supports the development of expert systems and their widespread application in industry and commerce. At a seminar held in 1985, it was reported that simple expert systems are now in practical implementation, with little risk and at relatively low cost, and are already producing modest gains. A whole range of expert systems has been suggested to help computer people. These include packages for fault analysis, training systems for operators and managers, operations advice systems, de-bugging advice for programmers, decision support systems, operations management and project management systems, intelligent non-numerical spreadsheets and improved project estimation.

As well as helping us to acquire knowledge, information technology systems will speed up the process of knowledge collection in a variety of ways. Electronic mail, electronic journals and text conferencing will support slower forms of exchanging ideas such as publishing journals, presenting papers at meetings and conferences and so on. In this way, it would be possible to replicate experiments or collaborate in international and cross-cultural studies in a much shorter time than is currently possible.

These potential benefits will, of course, only be realized through the use of technology to convey information. This is not a new phenomenon; forms of technology have been in use for a long time, but electronic technology gives us power and flexibility far beyond that of earlier technologies. It now includes, as well as computers, word processors, teletext and viewdata systems, satellite communication, cable television, video and synthetic speech production. The power and capability of these devices together is continually being enlarged and developed. However, whether we are truly able to harness the power of information technology to improve the quality of human life depends very directly on our ability to handle the human aspects of information technology. In some cases, these are often only too graphically described as the inhuman factors of information technology. There are now documented examples describing how frustrated users have physically destroyed their terminals or their microcomputers as a result of frustration with the system that they have been asked to operate. The following checklist published by the UK National Electronics Council in 1983 gives general, but nonetheless very useful, advice about how to introduce information technology effectively.

1. Planning the Development of a New System
New information technology will affect people not only after its installation, but also during its development. In order to ensure that people have a positive attitude towards it and are willing to contribute to its design, the following steps should be taken.

Setting up lines of communication
- Identify which people will be affected by the proposed changes: users, management, maintenance staff, people who have dealings with the users, and so on.
- Decide what information about the development should be given to these people, and how it can be realised in the most effective and timely way.

Planning the design strategy
- Identify those users who should contribute to the design of the system and the kind of contribution they will need to make.

● Decide what channels must be established to allow them to make their contributions.

Planning the transition
● Identify all the implications of the development and plan ways to absorb resulting deviations from normal routine with the minimum of disruption.
● Establish how much time will have to be spent:
  – by management assigned to controlling the development
  – by users contributing to the design
  – installing the equipment
  – training future operators
  – consulting with existing staff to inform them of the nature of the new system, the changes it will involve, and so on.
  – running old and new systems in parallel

2. Analysing the System
Standard analysis for a new system should investigate the following:

Potential users
● The names and job titles of all users of the system should be listed, including 'hidden' users, such as the company auditor. This list can then be used to ensure that all users are asked what they will require of the system.

Unattractive jobs
● Poorly designed jobs can result in dissatisfied employees and high levels of labour turnover and absenteeism. Such jobs should be identified so that they can be eliminated or improved on in the design of the new system. Similarly, it can be worth finding out which parts of their jobs users find satisfying and rewarding, so that these aspects can be included and enhanced in the new system.

Informal activities
● Over long periods of time, people tend to create informal solutions to infrequent but annoying problems in their work routines – by bypassing one or more levels of management, for example. In some cases the acceptability of a new system can depend on these informal activities being taken into account and catered for, even though they are not strictly part of 'official routine'.

User jargon and abbreviations
● Any jargon or abbreviations peculiar to the users should be noted and included in the design of the new system.

3. Selecting Equipment
All equipment should be as ergonomically sound as possible.

4. Helping to Design the System

Purchasers and users can contribute to the design process by considering the following:

Job design
- Those jobs which will be affected by the new system should be redesigned to provide interesting and varied work, the opportunity to learn, a sense of responsibility and purpose, future prospects, and a chance to gain recognition.

Monitoring performance
- People are likely to be more committed to their work if they can see how it has contributed to the performance of their organisation as a whole. Therefore, the system should be designed to collect and present them with such information.

Dialogue design
- That is, the construction of the content and format of communication between computers and their users. If users do not find it easy to communicate with their equipment, they will soon lose confidence in the system; therefore, considerable effort should be put into devising dialogues which are acceptable to the user.

Breakdown procedures
- Procedures should be designed, ideally, so that when breakdowns or faults occur, users are able to feel in control of the system by taking positive steps to tackle the problem. At the very least they should be told the cause and duration of the breakdown if this is known.

5. Workplace design

When information technology equipment is to be installed in a workplace, the following points should be considered:

Work station design
- The design and lay-out of the work station, terminal equipment and seating can significantly affect the user's efficiency. The work station should therefore be designed carefully and tested out by eventual users in a realistic context.

Lighting
- Glare should be eliminated as far as possible. Light fittings may need to be fitted with diffusers, windows with blinds. Uplighting should be considered seriously.

Temperature
- Some terminals emit excess heat, creating an environment that causes operators to feel uncomfortable and work inefficiently. All new terminal equipment should therefore be pre-tested before installation to make sure its heat emission is not uncomfortably high.

Noise
- Excessive noise can cause lack of concentration, irritability and stress. New equipment should be tried out in a room with noise levels similar to those of the room it is intended for, to see if it is distracting or disturbing. Soundproof covers or boxes can usually be fitted to printers, the noisiest type of equipment in common use.

Room lay-out
- Thoughtless installation of terminal equipment can result in a workplace which is cluttered and disorganised, difficult to move about in, and hard to clean and maintain properly. To avoid these problems the room lay-out should be planned in close conjunction with its occupants. If VDUs are to be installed, they should be positioned so that the user does not experience excessive contrast between the screen and the view around it, and also so that reflections in the screen are minimised.

6. Supporting users

Good support provides users with sufficient knowledge, skill and confidence to use their system efficiently and effectively. The following types of support should be provided:

Training
- Users should be given both general training in the way a system works and more specific training in the particular skills and knowledge they will need to use their system. Management should ensure that users can leave their jobs to receive the training they need.

Documentation
- User manuals should be easy to handle and use. The contents should be structured in a way that is logical and simple to understand. Jargon and unfamiliar terminology should be avoided. Users should have access to manuals as soon as the system is installed.

Secondary support
- When users cannot find out something about a system from their manuals, they generally turn to somebody else for help – an expert colleague, management or supervisory staff, or even a specialist adviser. Arrangements for this kind of support should be planned carefully, so that the appropriate personnel can be given the relevant training, and users should be told the kind of help they can get and how it can be used.

Planning for change
- Any information technology system will change and develop as the business changes and as inefficiencies and 'bugs' are discovered. Users who are encouraged to play a part in enhancing and maintaining a system are likely to view it as a useful tool and to be pleased by the chance to develop it further. The means for them to do this should be clearly laid out.

7. Maintaining the system

Equipment
- Regular checks should be made on terminals, work stations and workplaces to ensure they are providing the user with an efficient working environment. Where possible, it may help to give the responsibility for such checks to the users themselves.

Training new starters
- When new staff join the organisation, they should be given formal training in the system as soon as possible. Operating computer systems requires a great deal of covert mental activity, which newcomers cannot learn simply by watching experienced users.

Replacing local experts
- When local experts are about to leave, they should be asked to instruct a designated replacement in what they know.

Changing the system
- Users should be given appropriate training if changes which affect them are made to the system. Their manuals should be changed accordingly.

At a conference at the beginning of the 1980s, the then chief scientist at IBM said that the greatest obstacle to electronic information system progress is the failure of the more rudimentary systems of the past to reflect fully the relationships that people find comfortable and natural among one another and among their institutions. People and societies have had to adapt to the technology in order to enjoy its advantages. The major challenge of the 1980s is to reverse this process and to allow each person to define those relationships in whatever way he or she wishes and to reserve the right to change his mind. It is this challenge which faces project managers as they work to apply information technology solutions to business problems.

1.3 Selecting systems projects

There are many different reasons for the formation of a systems project team to develop a new applications system. Traditionally, these reasons have focused on the achievement of specific short-term goals related to the individual activities of the enterprise. In business organizations, these have usually been intended to increase the profitability of the enterprise by reducing costs or increasing revenues. Some typical examples are as follows:

(1) *To save manpower costs*

Reducing manpower costs in a line or staff department may be possible through the introduction of a computer-based system to carry out some of the activities of that department. An enquiry system, for example, that allows rapid access to data may enable a clerk to handle more transactions in the same period of time than was previously possible. In the 1960s, many new computer-based systems were introduced specifically to save manpower costs. Many large commercial organizations, and particularly clearing banks, automated many of their administrative procedures to avoid having to recruit new members of staff in times of labour market scarcity. This process is now exhausted as far as administrative staff are concerned, although the introduction of computer-controlled machinery and robots into manufacturing processes is now having a similar effect among blue-collar workers.

(2) *Improving customer service*

The introduction of a computer-based system may provide an improved service to customers by handling their enquiries and their transactions more promptly. In a highly competitive business, this may be a measurable improvement and, if the organization is dispersed, then the desired service levels may not be attainable without some form of data communication and perhaps the on-line processing of customers' transactions. This category of benefit might also include the rapid processing of customers' orders, the prompt despatch of spares or components, and the speedy resolution of customer queries.

(3) *Improving management information*

We have seen earlier in this chapter how the application of information technology seeks to improve the decisions taken by management. Many new computer-based systems are introduced partly with the justification of providing better management information so that better decisions can be taken. Examples of this kind of application could result in improved cost control, higher stock turnover, a reduction in work in progress, lower distribution costs, fewer production delays and better control of the sales and marketing function.

To suggest, however, that systems development projects arise in a haphazard way aimed at solving a random selection of an enterprise's problems would be incorrect. It is now more usually the case that a systems planning activity determines which projects shall be started according to the needs of the enterprise. The choice of which systems projects will therefore be carried out and consequently the range of applications on which project managers will work depends upon the business goals of the enterprise. Computer-based systems are developed to serve the overall information systems and physical systems that exist in an enterprise and are, therefore, derived from these. By implication, then, the systems planning function is essential to translate the business plans into achievable, developed computer-based systems. This process begins with the establishment of the business goals of the enterprise and the identification of the key result areas that it wishes to address. While these are very often profit, growth or market share – typical business enterprise key result areas – they may also be related to customer service, safety, staff development and so on. It is important to remember that many large enterprises dealing with government and the public service generally are not profit-orientated and have a range of key results which they wish to achieve that lie outside commercial parameters. In short, then, the key result areas specify the targets which the enterprise is aiming to achieve during some strategic planning period. From the top-level key result areas, divisions or functions of the enterprise develop their own specific key result areas. In data processing, for example, the management would be concerned with the development of new systems to contribute to the achievement of these higher-order key result areas. It is this process that leads to the selection and initiation of actual projects that will in their turn contribute to the achievement of the DP division's systems plans. These projects will all need to be reconciled to the DP division's standards that establish good systems practice within the enterprise and sound project development methods. This sometimes causes conflict when the achievement of a business key result area requires new or different practices to be implemented in the data processing department. We have seen, for example, how very many users have installed micro computers in their department quite outside the control of the data processing department in order to achieve their own departmental key results in a way that the DP department was not prepared to support. Similarly, it may not, for example, be data-processing policy to build prototypes for users or permit the use of query languages. Yet the urgent need to achieve a key business goal may require innovative computing methods to be employed. Figure 1.1 shows how this project-planning process can take place and produce a range of possible systems projects. A mechanism is then needed to rank this range of possible projects and to filter out those that are not worth pursuing at this time. Figure 1.2 shows the 'hopper' method taking systems project ideas in at the top and filtering out at the very bottom

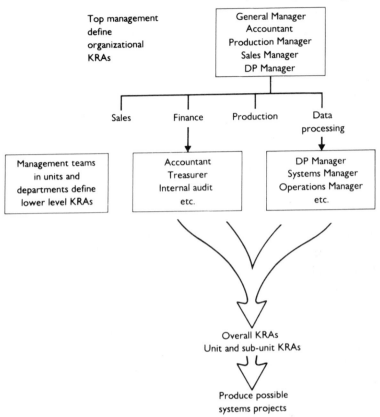

Fig. 1.1 Planning for projects

fully authorized projects. The first level of sieving, probably some committee such as a corporate systems policy group, rejects unsuitable projects immediately and lets through a range of possible unbudgeted ones which then compete for available resources. These then fall through to the pending area before a feasibility study is carried out for each of them in turn, following which the dead ducks are rejected and fully authorized projects begun.

1.4 The role of the project manager

The project manager's job is to carry through, to completion, the fully authorized projects. Consequently, he sets up a formal procedure to do

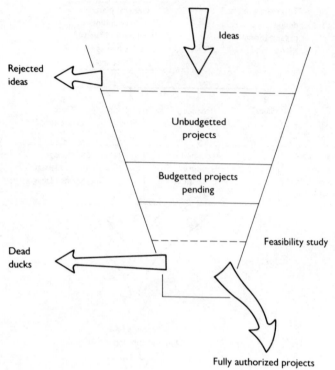

Fig. 1.2 Project selection

this. Later in this book, we shall see how the detailed project planning and control procedures are developed, but we do need to bear in mind that the amount of planning and control overhead applied to the development of a new application depends upon the size of the project, its technical complexity, the complexity of the application and the risk. While we recognize that all projects proceed through the same development stages, they do not all need the same degree of project management. The project manager will therefore need to decide on the most appropriate planning and control structures for his project. His success will be measured by how clearly the agreed product or service is delivered within budget, time-scale and to quality. Part of this success will depend on the organizational relationships between the project manager and the rest of the organization. In other words, it is important for the project manager to establish a firm base in the organizational hierarchy and, in particular, to be sure of his own upward reporting mechanisms. It may well be that he reports to a committee, a senior line manager or the manager of the department which will be the main user of the system. In any event, his success will be judged by that of the application development and therefore it is well worthwhile the project manager spending some time on establishing the criteria by which that success will be judged.

In Fig. 1.3 we can see the roles which the project manager will need to carry out. He will need to be technically competent so as to choose the technical computing strategies most appropriate for the application development. He will need to be an effective planner, a good controller of his team and sensitive to the problems of implementing change. He will need to be a good manager of his team and capable of training members to discharge their duties efficiently. Also, as we have just seen, he will need to be organizationally powerful so as to secure the implementation of the project and overcome any organizational obstacles that may lie in his way.

The role of the project manager can be divided into five components, as shown in Figure 1.4. It is likely that the project manager will need to review and constantly modify the business justification for the project.

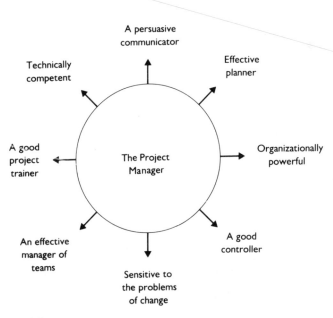

Fig. 1.3

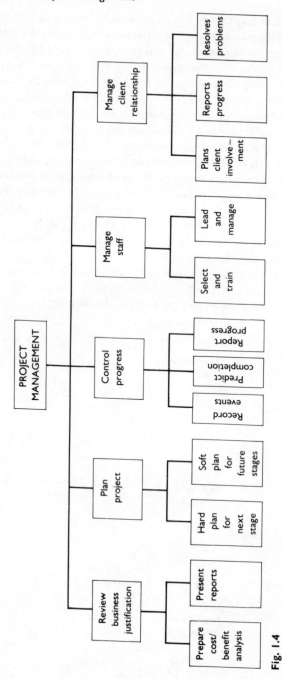

Fig. 1.4

Inevitably, as a senior systems person, he becomes involved with this at feasibility study stage and certainly in the preparation of costs and benefit analyses to support systems plans and requests from the users for their particular application development project to be put in hand. During the project development cycle, there is a constant need for everyone in the team to reinforce the benefits of the work being done. During the project, short- and long-term plans will need to be prepared in detail, justified and costed out; and the preparation of hard plans for the next stage and soft plans for the future stages will be an important part of the project manager's job. Controlling progress and recording that being made forms an equally important part of the project manager's job because there are always deviations from plan which, when identified, will cause re-planning to be done. Controlling progress means monitoring the work of others and hence there is a need to manage staff. We shall see subsequently in this book how the project manager can select, motivate and improve the performance of the members of staff in his project team. The project manager is also responsible for managing the relationships between the team and the client. This means many things – influencing the attitudes of the client staff, controlling and motivating his own people to present a positive and helpful attitude towards the client, selling new ideas, reporting progress and, in a service business, preparing cost information so that the client can be billed for the work that has been done.

Over-riding these five components of the project manager's job is the need to design a system to the client's satisfaction. This means not only the delivery of a working system, on time, to cost and within budget, but also a system that satisfies the user's social needs as well. Throughout this book the importance of involving users during the analysis, design and implementation processes and encouraging them to participate in the management of the project will be emphasized. Therefore, at a practical systems design level, it is necessary to take into account social as well as technical factors in the implementation of a new system. If a designed information system is to be truly effective and to provide the organization with cost-effective benefits, then it must normally fulfil a number of important pre-conditions:

(1) It must obtain the approval and the esteem of those who work with it and of those who use it. If a system fails to obtain approval before it is implemented, then most likely the implementation will be resisted, probably resulting in delays and in the subsequent misuse of the system. If it fails to satisfy the users once it has been implemented, it may be misused or supplanted by unofficial procedures. The fact that a new

system succeeds in obtaining approval before it is implemented, however, does not necessarily imply that it will also be approved once it is brought into operational use. Prior approval does not therefore mean that in practice the system will obtain operational approval.

(2) It must be capable of adapting or of being adapted to meet changing conditions and changing requirements. The designed information system that cannot be adapted quickly may affect the ability of an organization to continue in business or may cause its decline and eventual failure. It will certainly not enable the organization to adapt to changing environmental needs. However, since information systems are social systems and most social systems are robust, the informal human component often finds a way of overcoming the failure of the designed system, though at the cost of some inefficiency. We can see, therefore, that the failure of the designed system to adapt to change may result in it losing the approval of its users who devise alternative and unofficial systems to cope with the new environmental conditions. The designed system then becomes by-passed and falls into disuse.

(3) The various components of the information system – the designed and undesigned, the official and unofficial, the formal and informal – need to operate in harmony and be used to provide mutual support to the users of the system.

The project manager therefore needs to establish analysis, design and implementation methods that recognize these three pre-conditions and attempt to ensure that they are met. There are three things to be considered. First is the organization of the tasks to create a new system. Questions relating to the respective role of the professional systems analyst, designer and programmer and the role of the users have to be answered, as well as the best organizational structure to fulfil these roles. We have already mentioned the importance of the project manager establishing at what level he reports within the organization, and this is part of this organizational perspective. The second thing to be considered is concerned with the process of implementing change and with the tools and techniques available to help the process to be done efficiently and effectively. To some extent these are touched on later in the book when the management of implementation is discussed. The third is concerned with the process of designing the system so as to fit in with the capabilities, skills and cognitive preferences of those who will work with and use the system. This aspect of systems design is often called human factors engineering or ergonomics and is frequently described as making a system 'user-friendly'.

The project manager should therefore be concerned with establishing the criteria for good social design. If we begin with a very conventional view of how activities are organized for the design and implementation of a computer-based system, we can develop a model like that shown in Figure 1.5. This process starts with the setting up of a computer team to carry out the analysis and design for a new system and it shows very clearly the risks inherent in a traditional computer department driven approach. For the users, this produces:

(1) Low level of communication with the project team. There are a number of reasons for this:

(a) The user and the project team each have a very different image of the organization they serve and their understanding of each other's needs is restricted, which leads to inaccurate communications.

(b) The user is reluctant to reveal to the computer specialist the extent to which he is breaking company rules in the way he organizes his daily tasks.

(c) The user is afraid that information given to the systems analyst will be used against him later.

(d)) Because the user has little understanding of the design process, he doesn't realize what information is important and what is not.

(e) The user's principal interest centres around the tasks in his job and not around the information that he services as a by-product of his job. The computer specialist tends not to be interested in the user's job but only in that part of the job that is concerned with information flows.

(f) The inability of the user to visualize his behaviour within the proposed system. The computer professional is sometimes exasperated by the apparent unwillingness of the user to rule on the suitability of some new system component such as, for example, a new report.

(g) The practice of users to explain their actions in terms of rational behaviour when their actual actions may often be governed rather by intuition. Studies of managers at work, for example, have shown that there is a considerable discrepancy between how they claim to take decisions and how observations of their actual decision-making behaviour show that decisions are taken.

(2) Low understanding of the design process. In spite of the very widespread implementation of computer-based systems, computers are still very widely regarded as being esoteric and mysterious objects. This view of computers and of computer-based systems is reinforced by the way the media describe the functions performed by them. The phrase

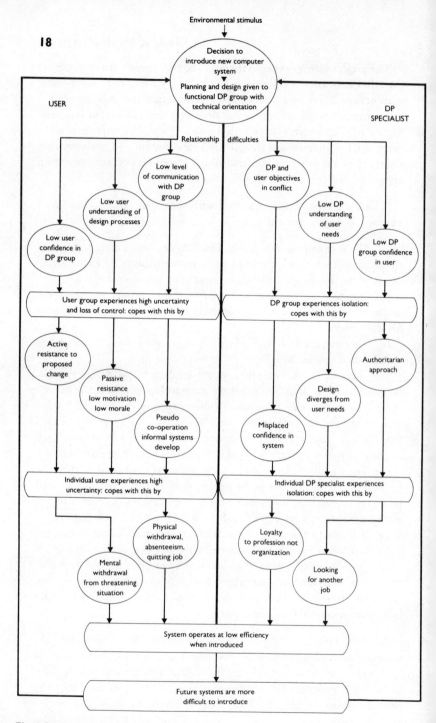

Fig. 1.5 The risks of the traditional approach

'the computer tells us' is used, for example, in otherwise quite sophisticated talks or articles about science or technology. The computer professional stands out within an organization as a person apart, perhaps endowed with a special understanding that enables him to communicate with computers. Although to some extent these difficulties have been broken down by the growth of personal computing and by the spread of computer education in schools, the user still has little real concept, in general, of what is involved in designing and implementing computer systems.

(3) There is low user confidence in the computer group. The computer group is seen as responsible for the disturbance and irritation which is a consequence of the development of new systems. Failings in the system once it is in operation are assumed to be the responsibility of the computer specialists. They appear to disregard many of the facts about procedures which the users think have been carefully explained to them. Overall, the computer group seems to have little interest in the individual problems of the user and, to many, is identified as a tool of management sent in to increase management control or improve productivity in the work situation. The users, therefore, are often observed as having little confidence in the computer group.

Conversely, for the computer professionals in the project development team, a parallel situation can develop:

(1) Computer specialist and user objectives are in conflict. Computer specialists tend to design systems that make for efficient use of computers and pay rather less attention to those that provide for human values and needs. DP staff have most often been trained in values that prize rationality and a scientific approach highly and that discount factors which cannot easily be quantified or treated technically. Many computer science and educational programmes have regarded organizational or human factor studies as outside their terms of reference, although it is encouraging to see that among the newer curricula being designed, this is becoming less true. Additionally, the DP specialist is not rewarded on the basis of user job satisfaction with the system but rather on the technical elegance of the system.

(2) Low DP understanding of user needs. The DP professional is handicapped in understanding user needs in a number of ways. He cannot be expected to be an expert in all business functions. His tools of analysis are confined to the examination of formal business procedures. Nearly all the methods that have been developed for systems analysis, such as the methods of data analysis, focus on a description of the

information flows within an organization and on the structure of the messages. Computer people traditionally see themselves in the role of journeymen with senses of values which tend to make them neglect organizational or social needs because their loyalty has tended to be towards their trade rather than to the organization that they serve.

(3) Low DP confidence in users. Many DP professionals, faced by users with different sets of values, with half-hearted co-operation, with reluctance to commit themselves to innovative and progressive ideas, lose confidence in them and tend to undervalue their intelligence and perception.

For the user group, the consequences of relationship difficulties with the project team are a high degree of uncertainty and a fear of a loss of control over the work situation. The group tends to react in ways which attempt to maintain or restore their control and reduce the uncertainty with which they are faced. They respond to this through either active or passive resistance to the proposed change, and through the development of alternative informal systems. Where active resistance is involved, the user group may, through its trade union or by means of a more informal grouping, such as a staff association, refuse to co-operate with the project team. Where the user is powerful, it can abort all proposed changes but where it is weak, sanctions may be employed to force the group into a more co-operative role. However, few systems which meet with active resistance will be implemented successfully. Passive resistance tends to be characterized by low levels of co-operation or by prolonged negotiations or discussions relating to the change by raising a whole variety of objections to elements in the proposal so as to slow down the work. Motivation in the group tends to drop and, with it, the efficiency of the work unit. A common way for the user group to regain control over its work situation is through the development of informal systems. These are often developed in parallel or may even replace the formal system and, in organizations which have developed networks of informal and often unauthorized systems, the formal ones tend to become degraded and ineffective because little user effort is put into keeping them accurate or timely.

When faced with these kind of relationship difficulties, members of the project team begin to feel isolated within the organization and respond to this by taking an authoritarian view or by designing systems on the basis of their own perception of the needs of the organization or in developing a misplaced confidence in their design. In this latter case, the absence of informed and constructive feedback from the users gives the project team a misplaced confidence that their design is acceptable to the user and workable in the long run. This situation progressively deteriorates to the point where all the adverse features shown in Fig. 1.5 are present. Then the new system will operate at very low efficiency when introduced, will need considerable maintenance in order to make it workable and will have cost much more to introduce than when first estimated. Within the organization, this can lead to a low level of confidence in computer-based systems generally, which will make it even more difficult to introduce new systems in the future.

Of course, the model described here illustrates the worst kind of situation, but nevertheless some of the circumstances described follow naturally from the perceived roles and needs of users and project teams. The DP specialist – the project team member – tends to gain esteem and will be judged by his peers on the basis of his success in introducing technical innovations. To further his or her career, there will be a tendency to push technology to its limit. By contrast, the typical user or perhaps a line manager can best succeed in terms of the achievements he is judged on, in situations of relative stability. So to further his own objectives, he attempts often to reduce uncertainty and to avert risk. Innovation, however, implies both risk and uncertainty. As an aside, it is interesting to note that the DP manager, in his role as line manager of the DP function, often takes on the risk-averting and uncertainty-reducing role when faced with innovations in the organization or techniques of his own DP department.

Participation by users in the analysis and design process is the one most powerful technique to avoid the isolation of the specialist and the loss of certainty and control by the users. It helps to strike a balance between the specialist's need to innovate and the line manager's requirement for stability. This is shown diagrammatically in Figure 1.6, which shows a contrasting situation to that in Figure 1.5. Here, users participate in the design process and, indeed, join in the design team, so that they share in the control of the planning and design process. As a result, conflicts relating to differing objectives are brought out into the open and an opportunity is provided for them to be negotiated and resolved. Users gain in confidence as they learn to cope with the problems of design themselves and have more confidence in the project team specialists because they are perceived as helpers rather than as opponents. Equally, the DP specialists lose their isolation and become more identified with the organization. Because users don't lose control and because they have a clearer vision of what the new system will do for them, then the extent of uncertainty is limited. It will still exist to some

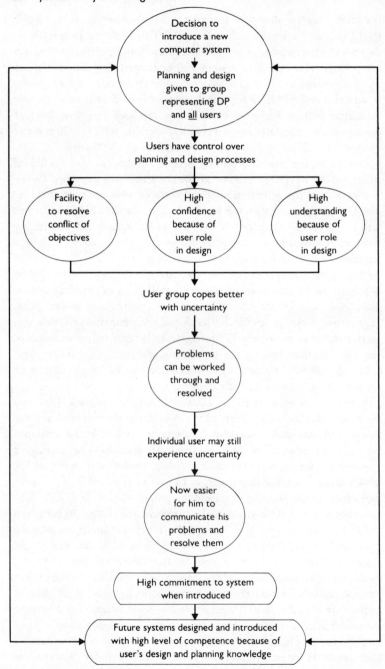

Fig. 1.6 A shared approach

degree, of course, because an innovative system cannot have precisely predicted outcomes, but the group of users and DP specialists together have an improved capability of communicating with each other and of resolving problems. Users are more likely to regard future changes as a worthwhile challenge because they themselves have played a part in setting the objectives for their system and in evaluating its design. Therefore, their response to uncertainty tends to be positive rather than an attempt to regain the *status quo* by means of counter implementation tactics. Above all, the users will tend to have a high commitment to make the system work and, by learning from the experience of designing systems, will achieve a higher level of competence to help them participate in improved designs for future systems.

1.5 How systems change takes place

It has been recognized for a long time that one of the key factors determining the ultimate success with which change is implemented is the actual process of introducing change. We have seen in the previous section how important is involvement by the user or the client in the analysis and design process. Since the ready acceptance of change makes life much easier for the project manager, it is important to know how change takes place and, of greatest importance, how the change is made; not what change is made, or the size of the change, but the change-making process. One well-proven model of the change-making process identifies three steps:

(1) *Unfreezing*. This is creating an awareness of the need for change and of generating a climate of receptivity to it.

(2) *Moving*. This includes developing new methods or learning new approaches, new attitudes and new behaviours to problems.

(3) *Refreezing*. This reinforces the changes that have occurred and stabilizes and maintains the new situation so that it becomes accepted.

In each of these stages are a number of key issues that need to be resolved; and it is important that they are before any attempt is made to move on to the next stage. These key issues, in their typical favourable and unfavourable manifestations, are shown in Figure 1.7. This particular model is now more than ten years old and has been applied to many systems projects. Although originally developed from experience of management science and operations research systems, the general lessons of the model are applicable in all systems development.

Favourable	Unfavourable

Unfreezing

Favourable	Unfavourable
1 Top and unit managers felt the problem was important to company. 2 Top managers became involved. 3 Unit managers recognized a need for change. 4 Top managers initiated the study. 5 Top and unit managers were open, candid. 6 Unit managers revised some of their assumptions.	1 Unit managers could not state their problems clearly. 2 Top managers felt the problem was too big. 3 Unit managers did not recognize the need for change. 4 Unit managers felt threatened by the project. 5 Unit managers resented the study. 6 Unit managers lacked confidence in the management scientists. 7 Unit managers felt they could do the study alone.

Moving

Favourable	Unfavourable
1 Unit managers and management scientists gathered data jointly. 2 Relevant data were accessible, available. 3 New alternatives were devised. 4 Unit managers reviewed and evaluated alternatives. 5 Top managers were advised of options. 6 Top managers helped develop a solution. 7 Proposals were improved sequentially.	1 Management scientists could not educate the unit managers. 2 Needed data were not made available. 3 Unit managers did not help develop a solution. 4 Unit managers did not understand the solution of the management scientists. 5 Management scientists felt the study was concluded too quickly.

Refreezing

Favourable	Unfavourable
1 Unit managers tried the solution. 2 Utilization showed the superiority of the new solution. 3 Management scientists initiated positive feedback after early use. 4 Solution was widely accepted after initial success. 5 Unit managers were satisfied. 6 Solution was used in other areas. 7 The change improved the performance of the unit.	1 Management scientists did not try to support new managerial behaviour after the solution was used. 2 Management scientists did not try to re-establish stability after the solution was used. 3 Results were difficult to measure. 4 Standards for evaluating results were lacking. 5 Top managers ignored the solution recommended by the management scientists. 6 Solution was incompatible with the needs and resources of the unit. 7 Top managers did not encourage other units to use the solution.

Fig. 1.7 Favourable and unfavourable forces at the three stages of change (S. L. Alter)

1.6 Summary

In this chapter, we have stood back from the day-to-day problems of managing the systems development process and considered some of the longer-term strategic and environmental issues surrounding project management. The systems planning process has shown how the development of new systems helps an enterprise to achieve its goals, and the importance of involving users has been emphasized. Our attention can now turn to the specifics of managing systems projects.

2 **Project development**

2.1 Introduction

The overall concept of a systems project can be summed up in simple global terminology: a payroll system, a weapons control system, an order processing system, a process control system, an air traffic system, a financial planning system. While each of these addresses different business applications, from commercial to defence, from general to specialist areas, the development process can be broken down into the same constituent parts. It is very easy to be led to believe that the more complex projects must be handled in a completely different way from straightforward developments that may need little time or resource. Equally, many people would argue that to develop a simple system requires little or no recognition of formal project stages. The project leader of complex or simple, large or small, indeed of any development project, should not be deflected by these views.

It is essential that the project leader first acknowledges the need for, and then recognizes, the specific stages into which every systems project is sub-divided. The penalty for not doing so will be an uncontrollable development process clouded in confusion and with misunderstanding between the end user and the development team. In these circumstances, time-scales and budgets will almost invariably be extended and poor systems developed. Repetition of sections of work will occur as progress to a later stage in the project will unearth areas of consideration which will differ from the original implementation. There will be a demotivation of the development team, and the user, if time is wasted reworking parts of the system which it was previously thought had been completed. It is an accepted fact of life that all systems will change during their development. This is perfectly understandable and manageable, provided it is controlled and is the result of genuine changes to user requirements, target hardware or software, or design considerations. It is not acceptable, however, if it has resulted from confusion during the development process.

Standards for the development of new systems vary, and development projects are broken down into different numbers of stages with different names. However, the basic building blocks are the same in all cases and are:

- Feasibility
- Analysis
- Design
- Programming and testing
- Implementation
- Post-implementation support

These are the major sub-divisions of any project (illustrated in Figure 2.1), and each must be formally acknowledged by the client or user before starting out on the development. Authorization will be required at each stage prior to moving on to the next, and we can see in the following sections of this chapter the significant aspects of each of these stages.

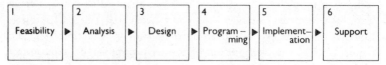

Fig. 2.1

2.2 Feasibility

In the previous chapter, we saw some of the ways in which projects start. Once this process has begun, we need to consider in more detail the practicalities of the proposed new system. We can expect the initiator of the proposal to be enthusiastic in its support, but a more balanced view is clearly required before a major investment is made in its development. A number of difficult questions need to be answered, such as:

- Is the envisaged system worth developing?
- Will the system improve efficiency?
- Will the system improve productivity?
- Will the system facilitate a reduction in staff?
- Will the system provide faster, better management information to enable better business decisions to be taken?

- Will the system ultimately save the enterprise money or make it more profitable?
- What will the system cost to develop and to operate and can it be justified?
- Will the system be acceptable to the staff using it on a day-to-day basis?
- How will the system affect the enterprise's organization?

These last three questions are considered later in Chapter 11. The shape of a proposed new system may well be influenced by the absence or presence of other computer-based applications. Constraints may be placed on the choice of hardware or software according to corporate policies about particular suppliers. In any feasibility assessment, these major policy considerations need to be evaluated as early as possible as they may well have an over-riding influence on the shape of proposed new designs.

In an earlier book (*Practical Systems Design*), the importance of conducting a thorough feasibility study and beginning the development process in a slow and carefully controlled way was emphasized. This point is reiterated here. Typically, it is the project leader's responsibility to carry out a feasibility assessment and he will, in consequence, have to live with the results of his work throughout the lifetime of the project. It is well worth making sure, therefore, that all interested people and departments have been involved in the feasibility assessment and that some measure of agreement has been reached as to the general direction of the project. All decision-makers need to be kept informed of progress on a regular basis, and the project initiator needs to keep in touch with the feasibility study on a day-to-day basis. The danger of too much involvement, however, needs to be borne in mind, and it is also worth remembering that at this stage we are concerned with a feasibility assessment only, and not an outline systems design. The objective of the feasibility assessment is to answer the question 'Is it worth spending money on a systems investigation for this project or is it clearly a non-starter?'

While the overall feasibility assessment will include an evaluation of the technical and social acceptability of the project, it is usual to find that financial feasibility far outweighs these other two considerations. Financial feasibility must not be interpreted simply as the cost in terms of hardware and manpower resources needed to develop the system. These are one-time costs which, while clearly playing an important part in the feasibility assessment, are only part of the total cost of the new system. The impact of the final system on the business must be assessed.

Typical questions that may be asked at this time are:

● Will the new system allow more work to be processed without an increase in costs?
● Will the client be able to reduce overhead costs and still maintain the same level of output?
● Can new product lines be introduced into the system more easily?
● Will better management control be available to enable the client to plan the use of the company's resources better than before?

A good feasibility assessment needs to be able to answer these basic business questions.

In short, the justification for any new capital outlay must be that it will increase the profit of the enterprise, improve the quality of service or products which are the business of the enterprise, or reduce expenditure. New systems developments will therefore be justified by cost and benefit criteria that ensure that projects which give the best return, according to the enterprise's policies, are those which are carried out first.

The assessment of technical feasibility is clearly based on preliminary outline systems design ideas relating to what can be accomplished with existing or imminently available technology. At times of rapid technological change, this is clearly a difficult process and it is not unknown for many systems developments to be started with only a hazy idea of the detailed technology that will be used to secure their implementation.

The assessment of social feasibility is assuming greater importance now than in the 1970s. This is due to the inroads that new systems are making on the work practices of end users and the growing realization among employee organizations of the need to negotiate new technology agreements with employers. This is covered in more detail in Chapter 11.

The culmination of the feasibility study stage is the production of the feasibility study report, which is presented to the client or user. The structure of a typical report is shown in Figure 2.2, together with an outline of the contents. The project leader will secure commitment to the proposal set out in the feasibility study report from the user or the client, and this commitment and the general agreement reached will be documented. Assuming, then, that the project is to go ahead, the feasibility study stage truly concludes with the preparation of the terms of reference for the next stage – investigation and analysis.

From the project leader's point of view, the feasibility study is probably the most difficult to manage as quantitative measures of progress are difficult to establish and to monitor. The data on which conclusions and recommendations are based is often scanty or speculative. The task of identifying and quantifying benefits in terms that senior management can readily understand and accept is difficult. Costings at this stage can only be considered budgetary and there is always the danger that they become embedded in management's minds so that subsequent phases of the project are judged against these early estimates rather than the more accurate costings derived as the project progresses. The project leader should protect himself from such constraints by including suitable caveats in his feasibility report and getting the necessary authorizing signatures.

2.3 Investigation and analysis

The investigation and analysis phase is concerned with two activities: the collection of information about the operation of the existing systems, and the identification of difficulties, problems and bottlenecks with those systems, together with the specification of fresh requirements which the newly designed system will have to fulfil. Terms of reference for this phase will have been provided by the client management and will typically include a statement of the objectives, the constraints, the resources available and the overall scope of the investigation and analysis activities. The project leader's first task will be to plan the investigation. A clear understanding of the scope of the project and of the resources available to him are obviously essential. Even so, the project leader may well find that this phase of the project is difficult to control tightly as many of the activities to be carried out are qualitative rather than quantitative. Being quite clear, for example, that all the information has been collected is difficult to do and yet essential. It may well be appropriate, therefore, to use one of the newer structured analysis and design methodologies to help in this stage. A well-established data-processing policy on standards within the client or user organization clearly eliminates the need for the project leader to choose from the range of methodologies available. However, the right choice of a methodology can be critical since their different strengths and weaknesses make some more suitable to one project than another. It should also be borne in mind that it is likely that some members of the team will need training in whichever methodology is chosen, either a refresher course or *ab initio* training. In addition, it is likely that the

Structure

Identification of Report
Management Summary
Introduction
Existing System Outline
Definition of Management Requirements
Review of Alternative Solutions
Proposed Solution
Development and Implementation
Back-up Resources
Action Plan
Conclusion

Contents

The front page should specify:

● the project title
● the project reference number
● the date of issue
● the authorizing signatory(ies)
● the distribution of the report

Management Summary

This should be a concise summary of the major recommendations of the report in no more than six pages so that it can quickly be understood by senior executives. It is usually placed at the beginning of the report. On very large projects it will not be possible to include sufficient detail on six pages and in these cases a separate document may have to be produced.

Introduction

This section should state the objectives of the report and of the project; it should refer to the terms of reference of the feasibility study and the constraints within which it has been conducted.

Existing System Outline

A succinct and precise definition should be given of the existing system using system flowcharts, tables and charts that show:

● the organization
● the scope of the system
● the main procedures
● the current staffing levels
● the operating costs of the system
● performance statistics of the current system, manual or automated
● a critique of the problems

Fig. 2.2 Feasibility study report structure

Definition of Management Requirements

This section should describe the user requirements of a new system, addressing the key areas identified above. Attention should be given particularly to operational deadlines, staffing levels and operating costs.

Review of Alternative Solutions

The major possible alternative solutions should be discussed and compared using clearly defined criteria. If possible, numerical weightings for each criterion should be established, but if this is done it is important to use the results from such weightings as guidelines only and qualitative criteria should be added to the evaluation as well.

Proposed Solution

More detail should be given of the solution which seems to best fit the user's requirements. This should concentrate on the effect on business operations, departmental organization changes, development and running costs, and benefits to the user department. Wherever possible, benefits should be quantified, and intangible benefits included as well.

Development and Implementation

A broad plan for the development and implementation of the proposed system should be given which identifies major activities, time-scales, resource requirements and costs. However, it must be stressed that this is an initial high-level estimate and that facts presented must not be held as binding because subsequent phases of the development will refine them.

Back-up Resources

To give the user department confidence in the new system, it will be necessary to state what procedures will be followed if, for any reason, the new system is unavailable through failure or breakdown. Alternative computer solutions should be given and, wherever possible, an ultimate manual standby. The restrictions of such standby facilities should be clearly specified.

Action Plan

This section states in detail the next development steps to be taken assuming that the feasibility report is approved. It should concentrate on the next phase – Investigation and analysis – but highlight major actions in all phases until completion. These actions should include those which are the responsibility of the user department as well as those of the systems development team.

Fig. 2.2 – *(contd.)*

project leader will set standards for his project in terms of documentation and progress reporting and these will be put into practice for the first time during the investigation and analysis phase. This clearly affects the productivity as the team members become familiar with the project standards and the required level of performance demanded by the project leader.

The main task during the investigation and analysis phase is the collection of data, and there is a great danger that insufficient time will be allowed for this activity with the consequence that future developments are built on inadequate foundations. The project leader's role during this phase is not to concern himself with the detailed techniques of systems investigation, but to prepare appropriate plans and make sure that the investigation proceeds in a controlled and orderly manner. A key characteristic of the good project leader at this stage is his ability to involve senior members of the user or client enterprise management, so that team members are able to proceed with their investigation activities without delay and that commitment from management to the development at this early stage is clearly visible to everyone in the client organization. The use of structured methodologies clearly facilitates this progress because of the need to involve users in the investigation and analysis process and, in particular, in the quality assurance work that validates the completeness of the investigation.

Conventional systems analysis requires the collection of many details of the current or proposed user system. The project leader must plan in detail the work of his team to ensure that every avenue is pursued and that sufficient time is spent with the user to collect all the data. This must then be examined and presented in a form which represents the needs of the new system. The standards for the presentation of this data should be defined, otherwise an incorrect document may be produced. However, it is far better to adopt a methodology from the onset of the project which will ensure not only a standard report at the end but consistent methods for the collection and reduction of data. This helps the user as well as the project team in that he knows how to prepare information for the interview session, being able to select only that which is relevant. It will be stressed throughout this book that it is essential to maintain close contact with the user department and to obtain its formal signed agreement to the various stages of development. The adoption of a methodology assists this significantly as it forces the definition of 'deliverable products' at an early stage and demands a far more vigorous review mechanism for use throughout the various phases of the project. This review mechanism must include the user at all stages.

Taking Structured Systems Analysis and Design Methodology (SSADM) as an example, during the feasibility study a set of data flow diagrams will have been defined for the current and proposed system, together with user options and process outlines. During the full analysis phase, the first task is to review the feasibility study – this is a mandatory review attended by the systems team and user staff. Adoption of structured walkthroughs for these reviews is highly recommended and consists of a review of a 'deliverable product', whether it is a data flow diagram, a system model or whatever, by a group of competent peers. Further recommended reviews occur at the stage of refining the current data flow diagram and at the creation of the required one necessary to provide the business solution. Such meetings enhance the understanding between user teams and project staff, especially as the reviews focus on specific stages of the analysis phase of the project and often on specific deliverable items as defined in the quality plan. Even if a formalized methodology is not used, it is essential that all projects have a quality plan stating clearly the standards to be used, the deliverable items and the relevant time-scales. This can then form the basis for meetings between user and project team.

Each methodology will have its own standard nomenclature for data and for the relationships between various data items. This clarifies to the development team and to the user the detailed operation of the existing system and of the one being proposed. The exercise of collecting data will be lengthy, but it must be thorough as it will form the basis for the subsequent systems design. The use of a methodology is therefore important as it imposes a discipline on the user and on the project team but still provides the tools to effectively exploit these disciplines.

It is imperative, should standard formal methodologies not be adopted, that the project leader ensures that some form of cohesive policy is maintained with respect to data collection, analysis and reporting. In this environment, there are a number of standards that address the needs of the analysis phase of the project. Equally, many commercial companies, especially systems houses, develop their own approach to systems analysis and the documentation associated with it. The deliverable at the end of the analysis phase is the functional specification, and this is illustrated in Figure 2.3.

2.4 Design

The end of the investigation and analysis phase will have seen the production of the functional requirement specification which defines the new system in terms of its business requirements. The purpose of the systems design phase is to specify the new computer-based system in terms of its technical content; the output will be a design specification.

Structure

Identification of Report
Management Summary
Introduction
System Overview:

- description
- data flow diagrams
- input
- output
- interfaces with other systems
- security and back-up
- special hardware or software

Cost/Benefit Analysis
Revised Development and Implementation Plan
Conclusion

Contents

Identification

The front page should specify:

- the project title
- the project reference number
- the date of issue
- authorizing signatory(ies)
- distribution of the report

Management Summary

This should be a concise summary of the major recommendations of the report in no more than six pages.

Introduction

This will refer to the objectives of the report and will reference the feasibility study and other relevant documents. It should highlight any differences in assumptions between this and the feasibility report. It should comment on, and explain, any differences in costs or benefits resulting from the analysis phase.

System Overview

This should contain a narrative of the new system backed up by comprehensive flowcharts or data flow diagrams. Input and output definitions will help to define the clerical user functions necessary for the satisfactory functioning of the overall business solution. Any interface to other systems, either manual or computer-based, must be identified and specified. The security of the proposed system should be expounded, together with any special back-up procedures. It may be that during the analysis phase a

Fig. 2.3 Functional specification report

need has been identified for some special hardware or software. It is in this section of the report that details should be given of such 'special equipment' and its impact on the proposed system explained, both functionally and financially.

Cost/Benefit Analysis

While still relatively crude, the cost/benefit analysis in this section should be a supplement to that given in the feasibility report. Analysis and programming estimates need only be to within $\pm 20\%$ unless better can be easily achieved. The distinction between development and live running should be highlighted in terms of computing power requirements and cost.

Revised Development and Implementation Plan

This will lay down, in the light of the analysis work done, the resources necessary to develop the system and subsequently to operate it on a live basis. A manpower plan should be presented showing resource allocation throughout the system's development and for the subsequent maintenance effort. It should define the best estimate to completion and signify the spread of the types of resource needed, complete with the identification of milestones to coincide with deliverables identified in the quality plan.

Conclusion

This section summarizes the conclusions and actions which have been generated throughout the analysis phase.

Fig. 2.3 – (contd.)

This mapping of the business requirements of the system on to the computer design is done at two levels. First, the logical design is developed which defines the user's outputs, inputs and processes. This is then mapped on to the computer system in terms of data files, program modules and supporting software to define the physical design. This physical design specification will also include the operational requirements, security levels and authorized user list.

The first step, however, is to understand the computing strategy of the client enterprise as far as hardware, software and operating systems are concerned. A declared policy on design methodologies, hardware and software will clearly influence the design strategy which the team is able to take. Similarly, development of the system as an independent, free-standing project, as opposed to fully integrating it with existing systems, will also materially affect the way design is carried out. The level of data and system security is becoming increasingly important and although very often a system overhead, the need for security can materially affect design considerations. Understanding these policy considerations and making sure that all members of the project team bear them in mind in their activities is a necessary overhead to the systems design activities. All policy questions need to be resolved at the beginning of the design stage, otherwise the direction of the work carried out may well stray from the desired course and major changes could therefore be imposed on the project team at a later date.

During the logical design stage, the project leader and his team will have frequent interaction with the users of the new system as it is imperative that a clear, concise and unambiguous definition of the proposed system is produced and agreed. Where the application is very complex and the development time very long, it may well be necessary for the project leader and the user project manager to agree that a definition of the ultimate target system can only be stated in outline and at a certain point in time. It may be necessary for both of them to agree that, as time progresses, there will be changes which will occur as the detail of the application becomes clear or as potential users change their minds about the outputs they expect from the system. Clearly this makes life difficult for the project leader, but it must be accepted that in projects that run over an extended period of development time, it is impossible to be absolutely precise at the outset what the final shape of the system will be. This means that careful change control is absolutely essential. Good documentation and regular agreement of actions at progress meetings during the design stage will make it possible for these changes, which will inevitably occur in the life of a long project, to be managed carefully. A major task of the project leader during the design stage will be to manage this change control process. This is not simply a technical activity since it involves agreeing changes and making accurate estimates of their impact on user management. Experience also shows that changes almost always result in increased cost and longer time-scales. A good relationship between the project manager and senior user management is essential if this is to be understood and accepted.

Once the enterprise's computing policies and the logical systems design have been clearly understood, the project leader can pay attention to the operating system facilities and the programming languages and techniques that are to be used during the development of the physical design. The selection of an appropriate development methodology clearly has an effect at this stage: it influences not only the design itself but the overall documentation.

Each methodology has its own nomenclature and standard layouts which are embodied in the documents produced throughout the analysis and design phases. It is therefore difficult to define the production of a design document in isolation as it tends to be an accumulation of 'deliverables' achieved throughout the life of the project. However, it may be useful to highlight the contents of a typical conventional design specification. This is shown in Figure 2.4 and serves to illustrate the areas of major concern during this phase of the project. The total analysis and design activities on 'typical' projects occupy around 30% of the total project life-cycle. Every effort must therefore be made to ensure the completeness and accuracy of information generated in these stages.

2.5 Programming and testing

The objectives of this phase are to produce the actual computer programs to handle the system data. The full range of programming skills and experience is required for this phase with senior programmers usually being responsible for the more complex modules or the control modules and less experienced staff responsible for coding the simpler routines. The choice of programming methodologies, programming languages and adherence to standards will be critical during this phase. Commercial systems have tended to be developed in COBOL, but the advent of fourth-generation languages will no doubt see a change to this. Real-time systems have typically been developed in CORAL or PASCAL but emerging government standards, both in the UK and the United States, will see a swing away from these languages to ADA, particularly for the military applications. Where microsecond timing is critical, such as in process control applications, it is often necessary to use ASSEMBLER level routines or even to write the whole application in ASSEMBLER. There is a large amount of historical information on the time taken to develop programs in these and other languages, and therefore the project leader should have a firmer base on which to plan his team's activities and on which to estimate the time it will take to produce finished programs. We should remember, however, that programming is still an art and is unlikely to become a precise science until automatic code generators take a major hold in the applications development scene. This means, therefore, that data from previous projects will need to be tempered with direct experience before developing estimates. Nevertheless, the project leader should find this stage of the project easier to estimate and control than the analysis and design phases.

The level of control exercised during the coding process will have a

(*Continued on p. 43*)

Structure

Identification of Report
Management Summary
Introduction
Objectives
System Description:

- the manual system
- the proposed computer system
- system flowcharts
- module design specifications
- print layout charts
- data file designs

Acceptance Description:

- test data specifications
- test data file design
- acceptance test design
- computing resource calculations
- system performance requirements

Documentation Design and Standards
Implementation Plan
Conclusion

Contents

Identification

This contains a basic summary of the document details as shown in Figures 2.2 and 2.3.

Management Summary

This is a brief summary of the document suitable for senior management to read. It is more difficult to summarize this phase of the project in a few pages than the earlier phases because of the amount of technical details necessary to be included in the full design report.

Introduction

This section defines the purpose of the document and its general layout. It may also embody the objectives of the report although sometimes these are contained in a separate section. It will also relate any preceding documents to each other and to this report. Significant milestones will also be highlighted in this introduction.

Systems Description

- **the manual system**
This section of the document outlines the manual procedures necessary to surround the new computer system, such as input generation, input methods and control, report

Fig. 2.4 Design specification report

distribution, audit documentation etc. The manual procedures for the changeover from the existing system to the new one will be described generally.

● proposed computer system

Here, a comprehensive narrative describing the system to be developed should act as an introduction to the more detailed flowcharts, data flow diagrams, program and file design specifications. It should describe the various components of the system and their interrelation. Major data files should be identified and significant reports described.

● system flowchart

A graphical representation of the system can be presented by producing a system flowchart. This will identify input, processes, output and files, and how the information flows between them. Using modern methodologies, this can be replaced by data flow diagrams. There already exists software to produce these diagrams from the original data analysis done during the analysis phase of the project. Greater automation will become evident with the progress of methodology development, and the automatic linking of these graphical outputs to subsequent phases of the project will become more common as IPSEs (Integrated Project Support Environments) become more widely accepted as tools of the trade.

● module design specifications

This sub-section of the design specification contains the 'meat' of the project. Each module is defined in detail with inputs, outputs and processes clearly stated, together with links from and to other modules.

● print layout charts

This section is self-explanatory, but the project leader should be aware that this is likely to be the most valuable area once the system has been produced.

● data file designs

Every piece of data in the system must occupy a defined location so that it can be easily retrieved for use in the relevant processes. The data can be held in discrete files with a small number of fields relevant to particular processes or it can be held in a 'database' using proprietary software for accessing the various data items. From a system development point of view, the discrete files are likely to cause the project leader fewer problems than an integrated database. This should be considered carefully when estimating time-scales and resources.

Acceptance Description

● test data specifications

In order that the project team can test all modules and the total system during development, it is important to formalize the data to be used at every stage. This will be representative of data used in the live system, but more stringent tests may be carried out on independent modules in the development cycle.

● test data file design

It is important to have test data available in a format that is readily accessible to modules and subsequently to strings of modules (link testing) with a minimum of reformatting. If careful consideration is given to this by the project leader, much time and effort can be saved in the closing phases of the project.

Fig. 2.4 – *(contd.)*

● acceptance test design

The design of acceptance test data should be a joint activity between the users and the system team. Both parties gain from this, with the users in particular developing confidence in construction of the new system.

● computing resource calculations

Often this is the most difficult area to define with any satisfactory accuracy. At this stage of the project, module and overall software sizing can only be an estimate. Path lengths can be defined and data storage evaluated reasonably accurately but the number of compilations and amount of change per module are very difficult to define. The amount of resource necessary during development will depend on the mode of development – batch or on-line – the phasing of activities, the development tools available, the language being used, whether a specific methodology is being used, and whether a proprietary database or other file-handling package is being used. Interaction with other users on the development machine should not be overlooked by the project leader when requesting computer resources.

● system performance requirements

Here the project leader must exercise strict control and influence over the user to ensure that only realistic performance criteria are requested of the system. The criteria should be matched to the use that will be made by the system. It is up to the project leader to negotiate with the user department to define achievable system performance parameters.

Documentation Design and Standards

If a recognized design and documentation standard is utilized, then it should be stated here and reference made to the relevant sections. If a methodology is used for development it, too, should be stated with extracts of the relevant sections.

If a new standard is to be defined for this project because of some unique characteristic, it must be defined in full in this section and the project leader must ensure that all his team members are conversant with it and adhere strictly to its code of documentation.

Implementation Plan

Throughout this design document, all aspects of the proposed system will have been defined. These can be broken down into many activities, and the project leader's responsibility is to pull together all these into a plan which can be implemented to ensure delivery of the software to time, quality and budget. It must also include all ancillary activities of the project team and user department to effect the full systems implementation. Training, data transfer, acceptance testing, parallel running, installation of new hardware and software, the writing of user manuals etc. must be defined and resources allocated.

Conclusion

This section provides a summary of the design document, the main aspects defined and critical areas needing special attention by the development team or the user department.

Fig. 2.4 – *(contd.)*

direct relationship to the amount of time spent on testing. Simple, clearly defined program modules will be easier to test independently and when integrated with other modules. In contrast, large, complex, convoluted programs which have not been broken down into smaller modules will be very difficult to control during the coding and testing process; additionally, they will be even more difficult to maintain once the system is live. It is essential, therefore, that the project leader takes a firm grip of the overall program design task so that discrete, controllable modules are specified for coding. This will make the task of achieving delivery deadlines much easier and will greatly facilitate the subsequent maintenance of the system.

It is important that sufficient time is allowed for the programming team leaders to manage their teams. It is easy to appoint a team leader because he is very good at programming in the language being used, and to expect him to do the coding as well as the management of the programming team. While this may be acceptable in very small teams of four or less programmers, once the team reaches five or more, the quality of the finished system is in danger if the team leader is not allowed to use all his time for managing and making sure the finished programs are of an appropriate quality. Very often, project pressures are against this as more 'managers' in the team will mean higher development costs. However, we do need to remember that this cost is more than recovered through the saving in implementation effort and through the subsequent ease of maintenance. It may well be that the project leader will need to fight for this programming management effort, but it is a fight that is well worth winning.

The testing phase can be broken down into the five major subdivisions of module testing, link testing, integration or system testing, project team acceptance testing and user acceptance testing. Provided that modules are small and clearly specified, testing can be accomplished relatively easily if the program structure has been well defined. Similarly, if the links between programs have been well defined then integration testing becomes relatively straightforward. However, if these links have not been accurately defined in the early stages of the project, serious problems will result and major time delays will occur. This should therefore be carefully considered when planning the program design. Project team acceptance testing and end user acceptance testing can sometimes become the same thing if, during the early stages of the project, an activity is established which lays responsibility on the user to define and to document the acceptance criteria that will be used before the system can be accepted. If it is planned that this information is available by the end of systems testing, it can be used as the project team acceptance test data. However, the progress of this data should be carefully monitored and contingency built-in to allow for the generation of test data from within the project team itself.

Module test data can be generated from a knowledge of the total range of data items at the input interface, the process stages and the precise output formats. If it is feasible, every combination of input data items, plus some spurious ones to generate error conditions, should be exercised. Recovery from the error condition should be tested to re-establish that the software is in a usable state. If the module is unusable, flags should be set to indicate to the adjoining modules its precise status. Link testing will demonstrate the ability of the adjoining modules to cope with any status flags set by the module under investigation. These link tests must be exhaustive in their testing of the interfaces between modules and hence a link test script will be very detailed and comprehensive, as illustrated in Figure 2.5.

An intermediary stage between link testing and system testing used by large projects is sub-system testing.

The sub-system test plan should address the following major areas:

- *External interfaces.* These are specific to the sub-system being tested, but typically include the transfer of relevant files, correct receipt of all valid transactions and record matching.
- *Major internal interfaces.* Such as the routines of a major enquiry in a large sub-system, and checking the action or occurrence of mis-routing.
- *Recovery.* To check that a sub-system re-establishes itself to valid status after an error condition.
- *Accuracy.* To check the correct application of rules specified in the design specification.
- *Flexibility.* To test the validity of software for multi-sites.
- *Documentation check.* To check that all documentation is complete.
- *Response times.* To check the actual times against those specified for this sub-system.

Beyond sub-system testing is the system test plan which defines those activities necessary to convince the project team that the system they have built meets the business requirements of the user specification. The contents of such a system test plan will include an overview of the role and scope of the system testing to be performed. Functional tests will ensure that business requirements are met, and final link tests will be

Link Test for AB

(a) Entry from a command page or any ready status display except format 26.

   (i)   Try AB out of context → 'INVALID COMMAND – NOT IN READY STATUS'.
   (ii)   Check that AB is not usable by supervisors with privilege levels 1 and 2 → 'INVALID COMMAND'.
   (iii) AB ⟨non-numeric characters⟩ → 'INVALID PARAMETER'.
   (iv)   AB ⟨enquiry number⟩ where the enquiry does not exist → 'INVALID PARAMETER – ENQUIRY XXXX DOES NOT EXIST'.
   (v)   AB 10000 → 'INVALID PARAMETERS'.
   (vi)   AB 1:23 → 'INVALID PARAMETER'.
   (vii) AB alone when user has no outstanding enquiries → blank page 1 of 1 (format 12.8).
   (viii) AB alone when user only has direct enquiries → blank page 1 of 1 (format 12.8).
   (ix)   AB alone when user has following number of indirect enquiries:

        1 → page 1 of 1
        2 → page 1 of 1
        20 → page 1 of 1
        21 → page 1 of 1
        22 → page 1 of 2
        42 → page 1 of 2
        43 → page 1 of 3

   Ensure that enquiries are listed 'oldest first' and line 1 is set to 'SUPERVSRY' and 'AB'.

   (x)   AB ⟨valid indirect enquiry number⟩ → command page and 'ENQUIRY XXXX DELETED' displayed.
          Check that the enquiry was successfully deleted!

(b) Entry from the command area of format 26.

   (i)   Check format context.
   (ii)   Test paging (next page, previous page and select page) → should be cyclic paging.
   (iii) Enter AB alone when viewing page 2 → should receive fresh display of page 1.
   (iv)   Repeat tests (a) (iii), (v), (vi) → same results.
   (v)   AB ⟨enquiry number⟩ where enquiry specified is not on the screen → 'INVALID PARAMETER – ENQUIRY XXXX DOES NOT EXIST'.
   (vi)   Repeat (v) with an enquiry that exists but on a different page to that on display → same result.
   (vii) AB ⟨enquiry number⟩ where enquiry specified is on the screen but no longer exists → 'INVALID PARAMETER – ENQUIRY XXXX DOES NOT EXIST'.
   (viii) AB ⟨valid enquiry number⟩ → current page redisplayed with specified enquiry removed and later enquiries shuffled up.
   (ix)   Repeat (viii) for a variety of situations, e.g.
        I Delete last remaining enquiry → blank screen.
        II Delete an enquiry which resulted in the next page being emptied → page total should change.
        III Delete enquiries from the last page → get previous page when last one deleted.

Fig. 2.5 Example of link test documentation

done under operational circumstances without the use of any development tools such as test harnesses. Response time testings will check normal and peak load responses and throughput tests will evaluate the system's ability to cope with the target performance expected from the system. Finally, forced system crashes will prove whether or not the system is resilient to failure.

2.6 Development support

Taking this in its widest context, it ranges from the provision of satisfactory office accommodation for the project team through to adequate computer develoment resources, team training and secretarial support right through to user training and system maintenance. Mention has already been made of the possible need to train the team members in new methodologies or software tools, and subsequent chapters deal in more detail with the training needs of the project team. The project leader will need to ensure, however, that training time and cost are included in the project plan and to allow for the impact on time-scales of team members developing proficiency in the new tools and techniques which the project will require. Clearly defined documentation standards must be available to the team, and the project leader will need to ensure that these are used throughout the project. It is his responsibility to get these established at the outset of the project; developing them as the project proceeds is quite unacceptable. Similarly, leaving the task of documentation to the end of the project – often a favourite approach of programmers – is equally unacceptable. It will be a lucky project leader indeed who finds that his team members actively like doing documentation but it is essential for the ongoing support of the project and it should be pointed out that to do it as the project proceeds is by far the easiest way, as opposed to leaving it all to the end.

The project leader will also need to ensure that adequate computing facilities are budgeted for and made available during the development of the project. At the simplest level, this may mean arranging testing time with the operations manager, but for new systems requiring the implementation of hardware, then the task to be carried out is clearly more complex. If the system is stand-alone and will be using its own hardware, then delivery of the hardware and software needs to be scheduled for the start of project development so that the same facilities can be used during development as will be used during live running. Unfortunately, this is a luxury often not afforded to the project team and they may have to share resources during the development process with existing operational systems. Close liaison will be needed between the development team and the operations department where new systems have been developed in this context alongside live systems. It will be necessary to plan well ahead and to make sure that adequate resources are free for the development team.

Access to key staff in the user departments will be necessary throughout the life of the project and it falls to the project manager to ensure that this is available. Structured methodologies already referred to will make user involvement obligatory during the investigation analysis and design stages, but there will also be a heavy load of user involvement during acceptance testing. It may well be necessary to train some of the user staff in the new system before it is complete. This can only be done by allocating team members to this task or by integrating members of the user department into the project team. This latter course is very much dependent upon the calibre and technical competence of the user staff and upon the commitment and motivation of the user department managers and staff to this course of action.

2.7 Summary

In this chapter we have touched upon some of the aspects of each stage of the development of a new computer-based system. All these stages lead towards the development and the implementation of a new system, and the project leader needs to be in tight control of all activities throughout this process. It is his responsibility and his alone to make sure that the work is done to time and within budget while adhering to the quality standards defined at the outset of the project.

3 Project planning

3.1 Introduction

The purpose of planning systems projects is to make sure that newly designed systems are delivered on time, within budget and to quality. This is not an easy task and there are many pitfalls along the way that make it difficult to achieve the goal. Some of these pitfalls will be predictable and, with the benefit of experience of previous projects, we shall be able to plan around them; others will be completely unpredictable, new and not previously experienced, and the skill of the project leader as project planner will be the deciding factor in bringing the project back on target. The project leader must therefore, be as prepared as possible to plan his way around problems that arise and to take remedial action so that the impact of pitfalls on project time-scales and budgets is minimized. Until relatively recently it has been almost the accepted norm that systems development projects will run late and over budget. However, modern planning techniques and better systems development methodologies are changing this situation. Additionally, more attention is being given to the early phases of systems development and to the need for setting sensible plans which can be executed effectively. In *Practical Systems Design*, the need for laying effective foundations to a systems development project were emphasized. The beginnings of a systems project are entirely creative: this does not occur through the earth moving or in a sudden flash of lightning, but is the result of discussion, consultation and contemplation. Problems unresolved here can never be satisfactorily eliminated later. It is, therefore, a slow process and while it may prove tedious, frustrating or even obstructionist to insist on a clear and widely accepted view of what is to be achieved, a slow start is a good start. Taking drastic action at the end of a project to tighten up and to control more and more of less and less of the work remaining is not the solution. All the effort is going into the wrong place. The solution is to have set effective plans at the beginning. This is akin to laying sound foundations for the building of a house; if these are skimped, if the concrete is too weak, if the foundations have been insufficiently dug, there will be a very real danger of the house collapsing. Likewise with the planning stage of a systems project: if it is skimped, there may be major areas which have not received sufficient attention and which, consequently, will have an adverse effect on the success of the project no matter how good the project leader and his team are. Many of the problems encountered in systems development projects will appear to be a function of the particular stage of the project, but they can often be traced back to inadequate planning in the early stages.

Why, then, do we plan? For some people, planning is an unpopular activity and, indeed, there are people who have no belief in it at all. It is said that smaller projects involving perhaps two or three people may quite easily reach a successful conclusion with little planning. This cannot be true, however, since an estimate of reliable costs and expected completion dates for systems development activities require prediction right from the beginning. For larger projects, without substantial planning it would be quite impossible to predict end dates and to judge whether development has been accomplished within time and budget. Without planning, we cannot estimate the number of days of development effort that will be required to complete the project nor make any determination of the manpower requirements in terms of the numbers of staff, their skills and their availability. Equally, it will be impossible to make any assessment of the elapsed time and to set target dates and work schedules so as to predict a likely completion date. Equally, the resource requirements of computer machine time, data preparation effort and so on cannot be estimated unless proper planning has been done.

Planning effectively is not an easy task. How do we know when a good plan has been produced? Is a good plan one that lasts a long time and is not subject to change? Any plan will have to be changed eventually, because replanning is an inherent part of coping with the changes that occur during the life-cycle of a systems development project. Any practical plan, therefore, must reflect these continual changes but, as experienced project leaders will know, this is difficult to accomplish. An experienced project manager describes some of the problems in project planning as follows:

One unfortunately common problem in project plannning is for senior users or DP managers to set targets and deadlines which might be motivated practically or politically. These dates may relate to company activities which reflect important business, legislative or corporate events and problem areas. The actual plans eventually created to justify these end dates are often produced by working backwards. It is therefore, not surprising that they may prove unrealistic although they may initially appear feasible. This type of approach often gives planning a poor image with demotivated staff not being consulted or committed. The more junior members of staff can easily become cynical about the value of planning and with management's attempts at planning.

As the following sections will try to show, good project planning prevents this from happening.

3.1 Planning process

Before getting into the detail of planning time-scales, team activities and computer resources, it is a valuable exercise for the project leader to be quite sure that the whole team understands the background to the project and its relevance to the business. While this may not be necessary for the project leader himself, since he is likely to have been involved with the project from the outset, it will almost certainly be necessary for some, if not all, the members of the team, since they are unlikely to have been involved in the feasibility study which led to the project being authorized. It is also useful to know who will be the main innovators and supporters of the new system in the user department. The project leader will have to work with all levels of staff in the user or client organization and it will be very useful to him to know where the centres of power and influence lie. It is well worth building up a good relationship with these key users so that their help will be readily available when difficulties need to be overcome. In order to ensure that formal control of the project is implemented in the user department as well as within the project team, it is very useful for a user project manager to be appointed. All communication between the project team and the user can therefore pass through the respective project leaders; a precedent that should be established from the outset.

An early task for these two project leaders is to collect as much data as possible from similar projects which may have been carried out in the client organization before. Inevitably, there will be lessons that can be learned from experience on previous projects, both by the project team and the users. The user project manager may also have information about how the business's main competitors deal with the business problems which the new project has been set up to solve. If company policy has already dictated the hardware and systems software which the new applications development will have to use, then a review of the generally available applications software may be useful at this time. These research activities are often overlooked in the haste to begin the systems investigation, but a small amount of time spent doing this kind of research will yield significant benefits later on. A decision also needs to be taken at this time about the development methodology which the project team will use since it will be the responsibility of the project leader to make sure that this is used efficiently. It may well be that the client's standard methodology will be used, and an assessment should be made as to its applicability to the needs of the particular application to be undertaken.

Finally, we should remember that planning is not an activity that occurs in isolation only once. It is an iterative process and the general ideas originally postulated in a first plan will need constant refining before a final workable plan is produced. This means that it will be necessary to go round the planning loop several times. The planning review may, for example, detect false assumptions having been made, impractical end dates predicted or an unreasonable number of staff having to be applied to a particular task at a particular stage in the project. It is highly unlikely, therefore, that the first plan will be the optimum one, and it will almost certainly be necessary for several attempts to be made before a workable plan is produced.

3.2.1 Major planning criteria

No planning can sensibly take place without an understanding of the framework and constraints within which the plan has to be accomplished. It is important, therefore, that the project leader establishes, documents and agrees with the user the constraints surrounding the plan for this new systems development project. He needs to establish the following eight points:

(1) What is the overall budget available for the project development?
(2) Within what time-scale must the system be developed?
(3) What is the expected implementation date?
(4) What staff will be available to the project?
(5) What hardware and software will be needed to run the live system?
(6) What hardware and software exists to develop the applications software and how close is this configuration to the one which will be used for the live system?

(7) What maintenance organization exists to support the system after it has been implemented?

(8) What commitment has the user department made as far as the availability of their own staff is concerned to assist in project development?

This is a daunting list of items to establish so early in the planning process but unless satisfactory answers are produced to these questions, then the framework within which project planning can begin will be unsatisfactory. There is no better time to get all the ground rules agreed and to establish the project leader's style than at the very start of the project. While there will be many instances when the project leaders will review the factors in this list, it is nonetheless important that project planning begins within an agreed framework.

The final project planning criteria to be agreed is the planning methodology itself. It is common practice now to plan in detail only the early stages of a project – perhaps up to and including outline systems design – but only to give an outline plan for the detailed design, coding and testing phases.

3.2.2 Activity planning

Having established all the basic criteria to his satisfaction, the project leader must get down to the detailed planning of the project, involving listing all the activities to be undertaken in each phase and documenting the interaction between them. At a first pass it is sufficient to list the activities, ignoring the time-scales that will need to be attached to them later. This list should be as comprehensive as possible and to this end a 'brainstorming' session with other project leaders or his manager is a good idea. During this session, don't be concerned about sequencing the activities, just list them. A flip chart or wall board is very useful in this process to record all the ideas, avoid repetition and serve as a memory jogger when compiling the final list. This activity listing should also be at a 'macro' level at this stage. Headings such as 'interview user staff', 'compile user questionnaire', 'define documentation' are sufficient. At a later date, they can be broken down to their constituent parts and time-scales allocated to them. It is important to remember the activities within the user department as these will, of course, form a part of the overall plan. Typical topics here will be 'review organizational structure', 'define acceptance tests' and 'plan operator education'. A checklist of general activities at this level is given in Figure 3.1; this is

Activity Checklist

FEASIBILITY

- establish budgets
- establish timescales
- interview senior user management
- get objectives of system documented and agreed
- do cost/benefit analysis and get user department agreement
- establish outline of proposed system
- produce feasibility report
- present feasibility report to senior management, especially including decision-makers for budgets

ANALYSIS

- design user questionnaire
- collect complete set of documents for existing system
- plan interviews with user department
- send team on interviewing techniques course
- analyse data flow
- document all findings
- define inputs and outputs of system in user language
- calculate volumes of data, both existing and projected
- document all findings and get user agreement
- present to user management
- define change control procedure

DESIGN

- define methodology to be used for design
- design input routines
- design processes
- design output reports
- define files
- design test and integration strategy
- document overall design
- define interfaces to other systems
- design module and program test data
- write program specifications
- user plus project team design acceptance tests

PROGRAMMING & TESTING

- check program specifications
- break down programs and modules to individual tasks and allocate to programmers
- produce code in chosen language
- define module and link tests
- ensure data available for integration testing
- check quality of code
- document all programs
- ensure user provides acceptance test data
- complete integration testing
- complete acceptance testing
- sign off system as meeting user specification

Fig. 3.1   Activity checklist

DEVELOPMENT SUPPORT
- define documentation standards
- define computer resources required for each stage of development
- decide program development mode: on-line or batch
- if on-line, define number of terminals required
- define the quality plan, building in milestones and measurable achievements
- define skills required for all stages of the project
- define training requirements of team
- identify inputs required from user
- define monitoring and control procedures
- set up progress meetings within team and with user department (do not forget to include ancillary departments which are relied on to contribute to the project)

Fig. 3.1 - (contd.)

not intended to be exhaustive but sufficient to get the project leader started in his planning. The larger the project, the greater will be the number of activities and the relationship between them. In this situation, therefore, it is probably wise to take main headings, then subdivide them into their constituent activities, otherwise a vast jumble will result.

3.2.3 Activity duration

It is important that the project leader is in full control of all project activities at all times. This will mean regular meetings, formal and informal, with the user department staff and his own team and regular liaison with the user project manager. These meetings will be most fruitful if there are measurable items of progress to be discussed; meetings coincident with major milestones are therefore most useful. Within the team, it is the project leader's responsibility to monitor his staff on a more frequent basis and this should be done by reviewing the tasks each individual has been assigned. The tasks, therefore, should be of sufficient length for the individual to achieve something measurable but not too long as to allow the project to drift.

The actual duration depends on the phase of the project but, in general, during the analysis and design phases, durations of four weeks are acceptable, while during coding and testing, two weeks is much more realistic. It may be sensible to set an activity such as 'interview all relevant staff in the order processing department' to be completed in three weeks, whereas another could be 'interview all relevant staff in goods inwards department' which might be completed in one week. It is therefore not essential to aim for tasks to be of the same duration, but reviewing is made easier if they are multiples of each other with a common end date. It should be stressed that these reviews must be conducted and documented with a list of actions resulting from them distributed to all team members.

In a very simple project, the overall duration may be just the sum of the activity durations. This can happen, particularly with microcomputer projects where one person acts as analyst, designer and implementor. However, the more normal situation is where there are a number of staff on the project, each with their own allocated tasks. The overall duration then depends on the interaction of these tasks and the sequence in which they can be done; this will be examined more closely later in this chapter. The number of activities which can proceed in parallel is very much a function of the number of staff available, but is always dictated by the phase that the project has reached; it is obviously nonsense, for example, to start activities in the programming phase before the analysis and design phases have been completed and accepted by the user as a true representation of the system to be implemented.

3.2.4 The quality plan

The question of quality will be examined in detail in Chapter 10, but the groundwork for a quality plan needs to be established at this stage. The main objectives of the project leader are to complete the project to time, budget and quality. In order to achieve the first of these, we have seen that it is necessary to plan the activities to achieve certain time deadlines; budgeting, which will be covered later, helps to achieve the second. To achieve the third, we need a quality plan so that we can actively plan it into the project and give it certain measurable criteria. If we do not do this, then the project leader and the client user will be unable to judge whether or not he has achieved his objective

The main attributes of a quality plan are a comprehensive list of 'deliverable' items with time-scales, project standards and a formal review procedure specifically aimed at quality. It is the project leader's responsibility to produce the quality plan and to adhere to it throughout the life of the project. There should, however, be an independent quality review board nominated in the plan; there should also be procedures for passing and failing specific aspects such as deliverables; and where failure is reported, there should be a procedure to deal with it.

The project leader should view the existence of this plan as a positive benefit to the project and the team. Regular independent reviews at certain critical phases such as start-up, end of design and integration can give positive feedback to the project leader which might not be apparent from within the project. Action can then be taken if necessary, to correct any slippage in deliverables, standards or performance. By constant reference to the quality plan, the project leader is perpetually reminded of the goals he is aiming for and this can only be of benefit to the project.

As with project planning in general, the size and amount of detail in the quality plan are a function of the size and complexity of the project being undertaken. For a small project such as writing a diary and bring-up system to be implemented in COBOL on one of the many mini systems available today, it could probably be contained on a sheet of A4 paper. In contrast, the quality plan for implementing a police command and control system on a multitude of minicomputers spread throughout the United Kingdom would run to several volumes. Indeed, some of the military projects which involve battle equipment, as well as computer hardware and software, require such elaborate quality planning and control that specific departments are established to deal with it. In these cases, there are quality standards which may be imposed by the user, such as the UK DEFSTAN 0521 for military projects and BS 5750 for commercial projects. In some instances these are mandatory, and in others advisory. The content headings for a simple quality plan are shown in Figure 3.2.

- Project name
- Project manager
- Quality manager
- Project standards
    - analysis and design methodology
    - documentation
    - coding
    - testing
- Deliverables
    - systems
    - programs
    - test specifications
    - documents
    - hardware
    - operating software
- Q.A. schedule and independent audits
- Testing procedures including acceptance test project plan

Fig. 3.2 Outline for simple quality plan

3.3 Planning tools and techniques

3.3.1 General

Throughout this chapter we have referred to making lists of activities, setting deadlines, defining tasks, formulating quality plans and so on. While these are essential constituents of project planning, they will be of little use to the project leader unless they are documented in a structured way which allows for easy comprehension and rapid updating. It is this representation which can be accomplished by using the various tools and techniques to be described in this section. Some of these tools have been available for a long time, such as bar charts (or Gantt charts as they are more properly known after their originator, Henry Gantt), but other more highly sophisticated project control and administration systems, commonly called IPSEs (Integrated Project Support Environment) are now becoming available. In between lie the network analysis tools with many variations from standard manual PERT charts through the various computer-based networks, each given their own proprietary names by their developers. The applicability of each method or product will be a function of the attributes of the project being developed, especially its complexity. Simple projects need simple planning tools, otherwise more time can be spent controlling the tool than the project. Complex projects with many interacting activities and dependencies will require the more advanced techniques which themselves will require a commitment of resources in terms of computer power and support teams. This, in turn, will have an impact on the overall cost of the project. It should also be borne in mind that a number of the newer analysis and design methodologies now being used have planning and control features built into them. It is essential to become fully familiar with the methodology to be used in the system development as it may well affect the project leader's choice of tools to be used later.

3.3.2 Bar charts

While these are the simplest form of planning tool, they can be used for many projects irrespective of application or even type of business. Their limitations are the number of activities which can be plotted physically on a sheet of paper in a comprehensible form, and the number of interactions, or dependencies, of activities. Quite simply, each activity is represented by a line whose length is proportional to the time taken to undertake it. Once all the activities are listed with their respective time lines, it is possible to see how long the overall project, or phase of the project, is going to take. However, great care must be taken not to start an activity which is dependent on a preceding one until that one is complete. A simple bar chart is shown in Figure 3.3. One serious limitation of bar charts is the difficulty of scheduling effort to the activities to ensure that the logical sequence is retained without overstaffing the project. The feasibility study in Figure 3.4 assumes a team of project leader plus one analyst. After setting up the project, the analyst is set to work defining and distributing a user questionnaire while the project leader starts interviewing the user management. As data is acquired from the interviews and the returned questionnaires, both team members can start to analyse it and draw conclusions. One can then analyse the potential savings from a new system for the application while the other can concentrate on the cost of developing, implementing and running it. Once the conclusions have been drawn and all the relevant information gathered, both team members can produce the final report to be submitted to their management and to the user's management.

The bar chart shown is very simple, but adequate for a project of that size and duration. It is easy to see how complex such a chart might become if used for a twenty-man project lasting two years. The only way it could be of use on such a project would be if the activities were at such a high level that they would be meaningless from the point of controlling any work.

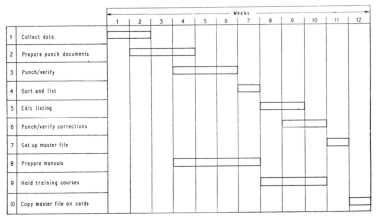

Fig. 3.3 A simple bar chart

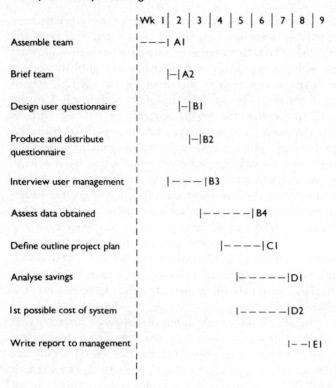

Fig. 3.4 Feasibility study bar chart

Control is exercised using bar charts by marking how far each activity has progressed at each project review meeting and comparing it with 'time now'. It is then easy to see which activities have slipped and the plan can be redrawn – remember, planning is a continuous exercise, not one done only at the start of the project.

3.3.3 PERT

The limitations of bar charts are nowhere more exemplified than where they are used for large projects. It was one such large military project, the Polaris Weapons System, which led to the evaluation of 'Programme Evaluation and Review Techniques' – PERT. Prior to embarking on this project, research had been completed on previous military projects and comparisons made between 'planned' and 'actual' time-scales and costs. The results showed the need for a tool to help with the planning and control of such projects, especially as the time-scale for Polaris was tight. PERT was born from a distillation of bar charts, milestone reporting systems and line-of-balance management techniques. It pays particular attention to controlling the time element and employs statistical analysis to evaluate the probability of meeting target dates throughout the life of the project; the production of progress reports are another major aspect of PERT systems.

In the initial planning phase of the project, it is vitally important that as many activities as possible are identified both by the user department and by the computer department. This will enable a comprehensive network to be drawn up and handled, especially if a computer-based PERT is available to the project leader. Associated with each of these activities is a time in which to complete the action and hence a prompt for all team members to think very carefully about the work they envisage the team doing throughout the project. The activities are then charted, each one starting and finishing at a node unique to that activity and connected by a line denoting the time to complete it. The interrelationship of activities which form part of a top-level activity can then be clearly illustrated. It may be difficult at the start of the project to define time-scales and hence, as a first pass to check that all activities are present, it may be ignored and the network established without time or resource requirements shown. This will allow activities to be moved around and different sequences tried to find the optimum. Figure 3.5 illustrates in network form the bar chart shown in Figure 3.4. It can be seen that activities B1 and B3 cannot start until both A1 and A2 are complete. Activities D1 and D2 can proceed in parallel but not until B4 is complete, although they can overlap C1. All of these activities must then be completed before E1 can be commenced, hence the 'dummy'

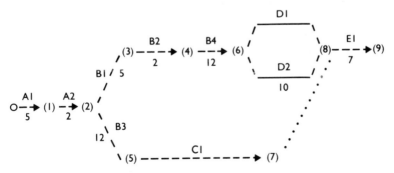

Fig. 3.5 PERT network

between node (7) and (8) which requires no effort but denotes a dependency by (8) on (7) being completed.

There are many detailed books published on PERT and PERT-like systems which can be used by the project leader. The strength of the technique at the planning stage is in its demand for a detailed definition of activities and the logical thinking required to link all these activities with accurate representations of sequences and interdependencies. This is most important and in fact should be undertaken by all project leaders whether or not they intend to use a PERT-type technique.

3.3.4 PROMPT

PROMPT is a planning and control system, developed by Simpact Systems Ltd in the UK, which recognizes two distinct organization structures within projects – management and technical. Much as we have broken down a project into various phases in Chapter 2, the PROMPT system breaks it down into six stages:

- Initiation, equivalent to feasibility
- Specification, equivalent to analysis
- Design, the same in both
- Development, equivalent to coding and testing
- Installation including acceptance testing
- Operation covering the early stages of live running, often referred to as warranty; this will be followed by a maintenance phase

PROMPT then recognizes the need for different skills in each of these stages and accepts that each may be managed by a different person – a stage manager. In practice, this can be one and the same person throughout the project but, nevertheless, it is important to recognize the different attributes required for the different stages of system development.

The top-down approach of this planning technique also uses the concept of 'the end product' where the completion of each stage signifies the production of a defined piece of software, hardware or documentation. This gives the stage manager an overall goal to aim for; he is then able to sub-divide this into smaller targets for the team members, each having their own end products.

The recognition of the need both for management and technical skills is taken further to encompass the interests of the business management, such as the board of directors, the user department and data processing department. A management structure is established, consisting of a project board, a stage manager and a project assurance team. The project board consists of senior managers from each of the three interested parties – business management, user management and DP management – and is responsible for the overall direction of the project. The stage manager is responsible for delivering the end products of the particular stage to time, quality and budget. The project assurance team has three responsibilities which cover all aspects of the project from beginning to end – administrative/financial, data and technical standards. Each has associated with it a co-ordinator who remains throughout the life of the project, ensuring the continuity and quality standards necessary for its success.

At first sight, the structure necessary to plan and control a project using the PROMPT approach appears to be relevant only to large projects, but this is not so. The most important factor is the recognition of the need for each of the functions; they do not all then have to be vested in different people. In small projects, many functions can come under the control of one person, but it is important to weigh up the effort required to implement the very formal procedures and documentation standards necessary for such a formalized approach.

The 'planning stages' of the PROMPT framework are initiation, specification and design. To help retain the differentiation between management and technical aspects of the project, there are plans for each with the management ones concentrating on financial matters and the technical ones on activities. In fact, management plans are broken down to four types:

● Project resource plan
● Stage resource plan
● Detailed resource plan
● Exception plan

Similarly, technical plans are categorized as:

● The project technical plan
● The stage technical plan
● The detail technical plan

These categories reflect those on the management side of the project, but management has one further plan – the exception plan – which spans both resource and technical aspects. The other plans are self-explanatory, but the exception plan is worthy of more explanation. It is required in any stage of the project where there has been an unacceptable deviation in the progress of schedule, cost or technical achievement from that expected. Overspends, excessive activity development times, major change requests should all result in an exception plan which defines the action necessary to correct the deviation. As with all PROMPT plans, it will contain:

- A graphic plan detailing technical activity and resource requirements
- A plan description explaining the deviation, the result if no action was taken, and the result of the proposed action
- Plan assumptions detailing any new business and technical assumptions made in preparing the exception plan
- Plan pre-requisites and risks, identifying areas of the project likely to be affected by the proposed recovery action

By standardizing on the contents of the plans, the stage managers are able to maintain consistency of documentation while producing detailed plans for action by the business management, user department and data processing department. All the major tasks of planning identified earlier must be tackled:

- identifying end products or deliverables and the associated technical activities to produce them;
- estimating the effort for each activity and the amount of resource needed;
- scheduling the activities on to the project technical plan;
- calculating the amount needed of each resource and costing it per month to prepare the project resource plan.

There are detailed procedures to control the distribution of all planning documents and to ensure that they are regularly updated. These are described in detail in the formal PROMPT documentation, but it is useful to summarize some of the facts obtained by analysis of past projects which show the breakdown of various projects to their six stages. This historical data in Figure 3.6 is useful when preparing the project resource plan and the project technical plan, as well as an indication of project resource and time allocation per stage generally.

We have only touched the surface of the PROMPT technique, but the project leader should see from this its relevance to controlling any project allocated to him. Unlike the simple bar chart and more complicated PERT planning tools which are of general use to any project, the PROMPT system specifically addresses the aspects of information systems.

Particular mention has been made of this technique as it has been

A. Resource usage per stage

Stage	% of *total* resource
Initiation	1– 3
Specification	11–20
Design	8–12
Development	31–51
Installation	18–28
Operation	3– 5

B. Elapsed time per stage

Stage	% of total elapsed time
Initiation	3– 7
Specification	9–21
Design	8–16
Development	22–54
Installation	12–28
Operation	6–14

Fig. 3.6 Resource and time allocation per project stage

accepted by many UK government organizations as the standard tool for planning and control of computer projects. This will have a direct effect not only on government employees but also on all the service companies that bid for government projects and who must train their staff in this approach in order to win the business.

3.3.5 Software development systems (SDS)

With the growth of information systems technology, the need to develop sophisticated software is becoming more and more evident. This software is often extremely complex in its operation and its development and, in consequence, the control of the projects and the level of detailed planning required is extremely high. The sheer volume of data and its interaction makes control difficult without a computerized technique. Accepting that such large projects break down, as with PROMPT, into technical and management aspects, there will be a need for staff addressing these two aspects to have access to much common data. It is logical, therefore, to hold that data in a project database and make it available to management, technical and administrative staff through a user-friendly interface with all the necessary security and access control measures. Such a concept was postulated as long ago as the early 1970s, and one particular development was pursued by Software Sciences and a government research establishment which resulted in a 'software tool' being produced and tested in the mid-1970s. Since that time, much experience has been gained by using this tool on live projects and a much upgraded system produced named SDS2.

The emphasis of this tool is to provide an environment for all data connected with a project to be recorded and accessed throughout the entire project life-cycle; this is illustrated in Figure 3.7. The data is

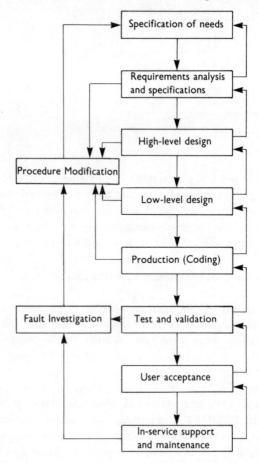

Fig. 3.7 SDS2 project life-cycle support

available to all relevant project staff but comprehensively protected against unauthorized access. Another important aspect of the tool is to make it readily available to the project team to use throughout the project. This is done using video terminals which allow straightforward access to factual data but which also have the ability to 'window' parts of the system to give the specific user his own view of that part of the database which is relevant to him. By affording this facility, the user is able to make his own changes to the data, which are then stored separately until he decides to apply them to the database and actually update it. This is very useful as the user is able to 'experiment' with different changes and view the results before he makes a final updating of the master database. There are, of course, built-in interlocks such that if two users wish to update the same data, the system suspends the application of the changes and prompts the users to decide how they wish to proceeed.

The hub of SDS2 is a network-structured database management package. Information is stored in records which are connected by a bi-directional link to form a network. This choice has been made from experience as that most applicable and efficient for analysis, design and general management use. From the user's point of view, this allows him to access records via links from both ends; it allows him maximum flexibility and does not impose any particular structure or methodology but allows him to define his own which is most relevant to the project. Access to the records is via a Data Manipulation Language (DML) which can search files, construct and manipulate intermediate tables of results, update the database and format the results ready for printing.

The product has many sophisticated attributes, but at the planning stage the aspect of most use to the project leader is the 'activity scheduler'. Needless to say, the same groundwork has to be done to use a tool like SDS2 as it does for PROMPT, PERT-type systems and even bar charts. Once the activities have been identified, they are expressed in terms of the position in the hierarchy, their duration constraints and dependencies. The activity scheduler will then perform a critical path analysis and produce earliest start and finish dates, latest start and finish dates, activity durations, sequences, and start or finish constraints. This is extremely useful, in fact essential, information for the project leader. Even more useful is the facility provided by the system by which each individual can update his own part of the activity network. By allowing each user thus to have his own private view of the database, the team leaders working for the project leader can update their activities, as can the programmers themselves. All these updates can be combined in the database and a consolidated picture accessed by the project leader. We will return to the control aspects of SDS2 in Chapter 5.

3.3.6 Integrated project support environments (IPSEs)

Like all 'production industries', systems development has progressed from simple beginnings with limited hardware, software and tools through to the current hardware with its vast storage and extremely fast processing ability, highly sophisticated software with many options and facilities available to the analysts, designers and programmers. Provision of all these facilities is not sufficient in itself, for without a structured environment in which to apply them, the system designer is lost. We have already seen the earlier developments of providing such an environment in PROMPT and SDS2 which typify a range of products from various suppliers. Current thinking is to extend the services provided by such products to encompass the total project life-cycle with detailed planning and estimation aids, full configuration control and documentation, quality records, and verification and validation tools. This total environment is known as an integrated project support environment, shortened to yet another acronym – IPSE. This in itself is an extension of specific project support environments such as APSE used in the UK development of the ADA compiler by the Ada Group Ltd.

While it is desirable for a user community to have one all-purpose IPSE to cater for all its system developments, there will inevitably be those best suited to commercial environments, or to real time systems development etc. By developing IPSEs in a modular fashion, it should be possible to insert new modules to cater for, say, different methodologies such as Jackson Structured Design (JSD), SSADM and Yourdon. The two main thrusts in Europe to develop IPSEs have come from the ESPRIT and ALVEY programmes. Probably the major significance of each of these programmes is that a link has been forged between the theorist operations in university environments and the business operations of systems houses and methodology suppliers to provide the user community with a tool based on sound basic principles, advanced systems thinking and yet practical enough to use to their benefit.

The Alvey consortium is headed by Software Sciences with Aberystwyth, Lancaster and Strathclyde Universities, with CAP and Learmonth and Burchett Management Systems providing the SSADM methodology. The nucleus of the IPSE being developed is Software

Sciences' SDS2 product described in Section 3.3.5, coupled with the SSADM methodology. The first brand product to be delivered by the Alvey group will be called ECLIPSE, previously known as PROMISE. This product will be developed around the UNIX operating system and will be available first on Digital Equipment VAX computers although it will obviously be ported to other manufacturers' ranges of machines. The product will support Ada, Pascal and 'C' with the MASCOT system design and SSADM.

3.3.7 The ESPRIT project contribution to planning tools

The ESPRIT program, like Alvey, consists of consortia of industrial and commercial companies combined with academic establishments. The objectives of this union are to put into commercial use the ideas generated in the academic units such that there is a positive advance in the fields of hardware and software engineering. Five key areas are of strategic importance to the ESPRIT program:

(a) Advanced microelectronics
(b) Software technology
(c) Advanced information processing
(d) Office systems
(e) Computer integrated manufacturing

This is much wider than Alvey, which concentrates primarily on software technology aspects. In ESPRIT, this area concentrates on developing efficient cost-effective methods for the production of high-quality software. Through this, it is aiming for a more rapid introduction of new products and a reduction of software costs in the overall project life-cycle. It is also aiming to bring more real compatibility between software tools and software production methods.

3.4 Budgets

3.4.1 General

It is the project leader's responsibility to complete the project to time, quality and budget. Time is often dictated by the user department in that systems may need to be completed by specific points in the annual cycle or to coincide with some re-organization. For example, it is not good practice to consider implementing payroll and accounts systems at the beginning of April in the United Kingdom as this is the finish of one tax year and the start of another when accounts departments are at their busiest.

Although quality will be discussed in Chapter 10, it is worth noting here that this is one of the most difficult areas of budgeting for the project leader. His only hope of achieving satisfactory quality is to allow sufficient time at the planning stage to plan for it in detail. As for budgets in general, the subject of this section, these are very much more measurable than quality as they can be monitored in quantitative terms which both project team and senior management can understand – £s. Budgets break down to two major areas, manpower and computing resources, and all other items can be amalgamated into these two. At the start of any project, the project leader may be faced with a situation where the time-scale is fixed and of paramount importance so that he has to establish the human effort and computing requirements to meet it. Conversely, he may be faced with a fixed budget for people and/or machine facilities and have to plan the project to complete in minimum time commensurate with the budget constraints. It is very important that the project leader establishes at the outset the budget guidelines that will be operated throughout the project. This involves getting agreement and commitment from both the user senior management and the computer department senior management. As with other major attributes of a system, it must be recognized that while an initial budget may be set, it will be subject to change, and control systems must be in place to manage this aspect. Remember that just as planning is a continuing process in the life of a project, so is budget re-estimation, and commitment must be gained at each change.

As his performance will probably be judged on how well he has controlled them, it is easy to think that budgets are the sole responsibility of the project leader. This is not true; they are the responsibility of every member of the development team and of the user department. Provided this is remembered, a responsible approach to the whole project will ensue and a much greater involvement will be felt by the team. It is easy to propagate this team spirit and involvement with small teams but more difficult with large ones. The subject will be covered in detail in Chapter 8, but suffice it to say here that clearly defined delegation in a large project is essential to make everyone feel a part of that project as a whole and responsible for their own particular sections.

3.4.2 Manpower budgets

We have already seen that the need for resources will differ depending on the stage of the project, as will the mix of skills required. In order to prepare a budget, it will be necessary to have as much information about each stage of the project as possible. Ideally, the project leader should only be asked to prepare such a budget for the next stage immediately following completion of the one in progress. He will then be armed with the best possible information on which to base his estimates. However, all too often it is the requirement of senior management that a cost for the project is established at the outset. This places the project leader in a dilemma as he will naturally wish to add sufficient contingency to cater for the unknowns which inevitably occur in the life of a project and yet he must not produce such a high cost that it is cancelled. From this aspect, the project leader in a software or systems house is probably mostly at risk as the trend is for clients to demand fixed-price contracts. Here, all estimates must be as accurate as possible and based on the fullest possible brief because commercial commitments will have to be made to the client, and upheld subsequently. A major difficulty in assessing the manpower requirements is, first, estimating the size of each stage at such an early point in the project and, second, knowing the calibre and skills of the team that will be available to the project leader.

Dealing first with the problem of project size, we will see later in Chapter 4 a number of approaches which can be taken to estimate the effort to complete each stage of the project. Once this effort has been determined in man days, the transformation to team size will be a function of the length of the project, recognizing that different types of staff will be required. For a relatively simple project, it will be possible to take each stage, divide the effort by the elapsed time and get a figure for the number of staff required. This will rarely be an integer so it must be rounded up if there is no flexibility in the stage length. If the overall project length is of most importance, it is possible to change the balance of the manpower needed by estimating all the stages and then doing a rationalization. For example, as a result of the effort calculations, the spread of the effort shown in Figure 3.8 is required to complete the project in an elapsed time of 88 weeks. Taking the raw figures results in not only fractions of staff but very high peaks during integration and acceptance which are not commensurate with the team size during programming and testing. So, without changing the overall time, it has been possible to smooth out this staff demand by observing the logical

Stage	Effort (man weeks)	Elapsed time (weeks)	Manpower Raw	Rationalized
Analysis	40	26	1.54	⎫
Design	60	24	2.50	⎬ 2
Program and test	100	26	3.85	⎫
Integration	60	8	7.5	⎬ 5
Acceptance	30	4	7.5	⎭

Fig. 3.8 Manpower calculation and rationalization

break in the project after design and before coding commences. By re-organizing the initial activities, it is possible to complete the analysis phase slightly quicker than originally planned and spend longer on the design without affecting the overall effort or elapsed time by employing two staff full-time. Similarly with the following three phases, it is not reasonable to nearly double the team size for integration and acceptance but it is possible to re-organize the coding activity to complete it slightly quicker and then spread out integration and aceptance with a consistent team size of five throughout. It must not be assumed that this exercise can automatically be done as there may be constraints on the program sizes and numbers, but in a straightforward project it is well worthy of consideration.

The project leader should always remember when converting effort and elapsed time to manpower that there are not 52 working weeks in the year – much as he may wish it, or need it! Allowance must always be made for holidays, sickness, training etc. A safe rule of thumb is to allow 30 man days for public and company holidays, 5 man days for sickness and 10 man days for training. Working on a 5-day week, this means 9 non-productive weeks and hence all calculations should be based on 43 man weeks, i.e. 215 man days of productive work. This means that elapsed time for the project equals man days effort $\times$ 52/43 (days). It is also very important that the project leader schedules company/public holidays and training well in advance to avoid clashes between these and significant project milestones.

At the initial planning stage, the project leader is unlikely to know precisely the structure of his team and the specific personalities to work on the project. Hence he must prepare his budget based on 'the average man'! This alone illustrates the need to re-plan on a regular basis to cater for people's actual performance as opposed to their assumed ability.

In a 'small' project, it is likely that the project leader will do the majority of the estimating himself, but this will not be the case in larger projects where there is a second layer of management, probably called team leaders. Here the project leader has a double problem – assessing the reality of figures put forward by the estimators as well as the potential of the team members. Only past experience of the team leaders' ability at estimating will help the project leader to resolve this dilemma and produce a reasonable budget. Secondary, or check, estimates are very valuable if the organization is of a sufficient size to have a pool of experienced estimators.

A typical manpower budget sheet for a project is shown in Figure 3.9. This is completed at the beginning of the project and enables the project leader to plan and cost his people resources, not only in man-effort but in cost by applying the cost per resource for each of the grades of staff involved. In a user environment, this cost is usually obtained from the financial department and consists of a person's salary, the cost to sit him at his desk and a contribution to company/department overheads. In a software house it is the charge-out rate for the particular grade of person on the project. The form is quite self-explanatory and easy to upgrade to record the 'actuals' in labour and subsistence once the project is in progress. A completed plan form is shown in Figure 3.10 giving a month-by-month breakdown of the total manpower budget including the subsistence relevant to the project team. This is a general form applicable to any size of project; continuation sheets can be appended as and when necessary to allow for more staff and for larger projects.

3.4.3 Computing resources budgets

In the last section we discussed manpower budgets and touched on the difficulties likely to be encountered when developing these budgets because of the variability of people's skill levels and ability to estimate the size of tasks to be undertaken. With computing resources, we do not have to cope with such variability as the resource being used can be defined in precise units. A CPU second is always the same on a specific machine no matter how 'young' or 'old' it is, a block of disc space is always a block of disc space, line printer output rates are always consistent given the same environment. So, with some of the major variables removed, surely it should be possible for the project leader and his team to estimate the amount of machine resource required fairly accurately and hence establish a very manageable budget. In the next chapter, some methods of preparing such estimates will be proposed, but

Manpower budget summary   Page .... of ....
Project name
Project
Reference   Date
Project
Manager   User Project Manager

Name	Grade or title	Days effort / Cost/day	Month 1	Month 2	Month 3	Month 4	Month 5	Month 6	Month 7	Month 8	Total
		Days									
		£									
		Days									
		£									
		Days									
		£									
		Days									
		£									
		Days									
		£									
Monthly labour cost											
Cumulative labour cost											
Monthly subsistence											
Cumulative subsistence										Total £	

Fig. 3.9 Manpower budget summary

Manpower budget summary   Page .... of ....
Project name
Project Ref
Date
Project Manager   User Project Manager

Name	Grade or title	Days effort / Cost/day	Month 1	Month 2	Month 3	Month 4	Month 5	Month 6	Month 7	Month 8	Total
P. Leader	Proj. Man.	Days	23	22	10	21	15	20	10	5	
		£	5750	5500	2500	5250	3750	5000	2500	1250	31500
A. Llyst	Sen. Analst.	Days	23	22	20	10					
		£	4600	4400	4000	2000					15000
S. Prog	Sen. Prog.	Days			21	20	18	15	22	10	
		£			3150	3000	2700	2250	3300	1500	15900
P. Gramm	Pro-grammer	Days				21	20	19	21	10	
		£				2100	2000	1900	2100	1000	9100
J. Mer	Jun	Days				21	15	22	19	10	
		£				1470	1050	1540	1330	700	6090
Monthly labour cost			10350	9900	9650	13820	9500	10690	9230	4450	77590
Cumulative labour cost			10350	20250	29900	43720	53220	63910	73140	77590	
Monthly subsistence			560	290	150	720	600	400	510	280	3510
Cumulative subsistence			560	850	1000	1720	2320	2720	3230	3510	
										Total £	81100

Fig. 3.10 Manpower budget summary – example

experience shows that machine resource budgets are even more difficult to compile than manpower budgets.

The variability in manpower skills and performance are reflected in machine usage. The careless coder will make many silly mistakes which can result in numerous compilations before getting the program right. If the development is being done in a batch environment, some of these may be trapped by desk-checking, but with a very strong bias to on-line program development it is more likely that the programmer will work away at the terminal and leave it to the machine to find his errors. There are mixed views on the merits of full on-line development with programs keyed directly into a video terminal compared with a degree of desk-checking or even full batch development. It is not the subject of this chapter, or even this book, to discuss the merits of these development environments, but suffice it to say that they can be a major influence on the computing resources required. The effect of this influence can be hidden if the project team is fortunate enough to be given the target hardware for their sole use in development, such as in turnkey systems. However, this is a luxury rarely afforded to development teams who usually have to compete for machine resources with live running systems and/or other development systems. The difficulty in providing reasonable budgets in such a competitive environment is caused by the unpredictability of the pattern of use of the machine by all users – batch development or on-line – and the interaction of these different patterns of work.

It is easy enough to account for the machine resources after they have been used. There are many different formulae for doing so, ranging from simply recording the number of hours development terminals are logged on to the machine and multiplying by a cost, up to very sophisticated formulae involving cpu usage, disc space, disc access, magnetic tape use, line printer output volumes and terminal connect hours etc. In practice, it is wise to capitalize on this past history of machine usage during similar project developments and use it to predict current resource requirements.

Included in the budgets must be the physical requirements as well as the computer usage – how many terminals of what type for how long, what local or screen printing facilities are needed, are separate exchangeable discs needed for the development etc.?

All these aspects of computer resources must be carefully planned at the outset but, as with manpower estimates, they must be reviewed and updated on a regular basis. Monthly is usually sufficient for this but any significant change in project team levels or major change of direction of

Machine : _____

Operating System : _____

Requirements	P	P	P	P	P	P	P	P	P	P	P	P
Terminals and connect hours per day (N × hrs/day)												
Average disk storage (KB/day)												
Mag tape drive usage (hrs/day)												
Printing requirements (pages/day)												
Batch queue usage (hrs/day)												
Standard software												
Special software												
Special hardware												
Special security measures												
Special facilities, e.g. dedicated use												
Required operating hours (outside 0900–1800)												
Others, e.g. Demos												

1. This form should show a realistic estimate of your computer requirements for the next 12 periods
2. Peaks should be marked * and details given overleaf
3. Add any one-off target date overleaf

Signed _____

Fig. 3.11 Computer resource budget

the project should precipitate a review. Likewise, changes in operating and system software, development tools and machine loadings could necessitate an update to the budgets.

Figures 3.11 and 3.12 show two typical forms for presenting computer resource budgets. Figure 3.11 is biased towards the development team which has little experience of estimating or limited access to historical data for comparison. Figure 3.12 is biased more towards a detailed estimating environment which attempts to predict cpu usage for on-line and batch development loads. These forms can be amalgamated and adapted to suit the specific development environment and are presented as examples rather than definitive standards.

As technology moves forward and the development environment, as well as the live systems environment, changes, there will probably be a simplification in the production of computer resource budgets. Trends towards individual workstations will mean counting the number of programmers and multiplying by the cost of a workstation to get a start-up budget. This can be amortized over the life of the project if the units

Computer resource requirements

Project name		Project number	
Contact name		Start date	
Time span (mths)		Number of staff	
Machine type		Operating system	
Languages & versions		Software tools & versions	
Special equipment		Work profile	

Special processing requirements

Consumables required	Working set requirements
Signed	Date

Fig. 3.12 Computer resource budget

	Machine type	Operating system
Period		
Week		
Interactive (hrs/week) — Connect		
CPU		
Batch (hrs/week) — Day		
Night		
Disc store/1000 blocks		
Paper use No. of sheets — Standard		
Others please specify		
No. of terminals required		
Machine load estimate (L, M, H) low, medium, high		

Any other notes or comments

Fig. 3.12 – (*contd.*)

are bought specifically for it or accounted for at the agreed company charge per annum based on a write-off period, usually of five years. There will still be a need for central resources to back-up these workstation networks but the central facilities should be simplified to large data storage and file movement between centre and workstation.

3.4.4 Summary

In this chapter we have attempted to set the scene for the need to plan. It is a vital activity at the start of any project but must not be considered as a once-only activity. Regular, constant updating of the information is essential to reflect the current status of the project and to predict the forward need for resources, be they labour, computer facilities, space, desks etc.

Examples of how planning can be effected have been illustrated from the simple bar chart through to the all-embracing integrated project support environment. The latter may seem a pipe dream but real effort and money are being poured into generating such software and there are very positive signs that such planning and control mechanisms will be an ever-increasing pre-requisite for companies tendering for large government contracts. The need to manage configuration control, ensuring that all aspects of the software generated keep in step, and to ensure that comprehensive documentation is maintained are only two of the facilities provided by such systems. The ability to control the software development programme and make up-to-date decisions on information common to all project members can only benefit the quality and timeliness of the ultimate deliverable software.

However, we may have to wait a few years before such systems are fully available and then their use will have to be carefully selected. The project's size must be such that it can support the necessary staff to manage such a facility: justifying the existence of this team is often more difficult than writing the software, as senior management have yet to be educated that this is not just another overhead but a real productive tool to aid the company in successful project management.

The information technology environment is probably the most rapidly changing area of business, with more and more processing power being made available to users at lower and lower prices in ever-diminishing space. This means that even the small development projects can justify the use of some form of computerized planning and control system. Simple spreadsheets are prolifically available on personal computers which, if nothing else, allow the project leader the ability to plan and re-plan in minimum time with consistent accuracy. With the growth of workstations capable of multi-function working, there can only be an increase in the use of project control tools and there are already many such packages available based on PERT-type networking. This must ultimately do nothing but good to the industry and result in better-quality systems being delivered to end users.

4 Estimating

4.1 Introduction

Estimating mantime and machine resources is still as much an art as it is a science and, consequently, it relies substantially on experience. New project leaders therefore invariably get it wrong. Developing a 'package' for subsequent tailoring and implementation at many sites is relatively straightforward and estimates can be based on statistics acquired during the initial development phase. Unfortunately, the majority of project leaders don't have the good fortune of leading such simple projects and will, on the whole, be dealing with new ones, bespoke to the application required. If however, he works in a well-organized DP department, then there should be detailed records kept of previous projects that can be used as source data on which to base estimates for the new system developments. However, great care must be taken not to regard these as entirely relevant as every system has unique characteristics making it different from its predecessors. The degree of difficulty in estimating varies with each phase of the project. The feasibility study is often constrained in time by senior management whereas analysis and design phases vary enormously in time and resources. Once the programming phase has been defined, however, and the project leader knows how many modules of each complexity are to be programmed, then average coding rates can be used to determine the number of man days per module. The standard of programmers varies widely and hence any figure of 'number of lines coded per day' must be assumed to apply to the average programmer. The variability of the standard programming skills in the team which the project leader eventually controls is only one good reason why the project effort and time-scales need to be re-estimated on a regular basis, at least once a month. We shall see in Chapter 5 the need for a change control system to cater for changes in requirements which were not identified at project start-up and which arise during its development cycle. This procedure can, however, also be used to cater for major problems encountered because of estimating errors, but only with the full co-operation of the users as it will have an impact on their time-scales and budgets.

A variety of techniques have been evolved to estimate software development costs, ranging from simple rules of thumb to extensive models of the proposed system. Like planning and controlling projects, the project leader should select the method best suited to the project and to the quality of the data. It is no use spending many hours computing highly complex formulae if the data is very superficial. A number of these different techniques will be described later in this chapter.

Estimating machine resources – often referred to as 'sizing' – can be quite accurately assessed by knowledge of data paths and overall data storage requirements, but predicting the actual performance of the system is much more difficult. This is particularly so if the application is to run in a shared environment with other applications and developments. For projects developed within a 'user' company, the criteria for estimating are usually based around minimum cost and acceptable time-scales. There will obviously be a lot of in-house discussion both on time and cost, but at the end of the day the decisions will be made within the company. Where systems are being supplied by an outside organization, such as a systems house, the estimating view of software as seen by them will have further constraints. In most such instances, there will be competition to win the order from the customer and hence there will be pressure on the estimators to produce a least-cost solution. Several re-designs may be made to achieve this goal, and some 'marketing decisions' may need to be made with respect to costings. However, such decisions should be taken consciously and the project leader should be allocated the 'real' budget.

4.2 Estimating manpower resources

4.2.1 Tasks for preparing estimates

In order to prepare any overall estimate for a project, it is necessary to undertake the four major tasks below:

- *Assess the project.* What is to be delivered?
- *Identify all the activities.* Break down the project into as small units as possible and link them together to form the whole project.

- *Evaluate the net resource requirements.* Define how much of each resource is required and who will provide it. This should exclude contingencies and overheads
- *Cost the resources.* This is very relevant to the user department, but often the project leader is primarily interested in people and machine resources.

Time must be allocated to undertake these tasks thoroughly. If an accurate estimate is required, a lot of effort will need to be expended and as many cross-checks as possible built into the exercise. If the project leader is able to control this estimation, he is in a strong position to influence the final estimates within which the project is developed. This is particularly important if he is to be judged formally on whether or not he meets these targets. Some organizations reward project leaders by incentive bonuses against them meeting time, quality and budget criteria. In this instance, if the project leader has compiled the figures, they must be checked by an independent authority as it is only human nature to set 'easy' targets, especially when money is involved. Similarly, if he is given someone else's estimates to work to, he should check them out as he may have been given an impossible schedule to work to.

As a general rule, estimates should be done by at least two independent people whenever possible. If they are within 10–15% of each other, the chances are that both are correct and minor divergencies can be rationalized by discussion between the parties who prepared the estimates. If there is a major divergence between them, it is best to disbelieve them all and go back to different estimators to get a new perspective on the project, briefing each of them to ensure thoroughness of work and explicitly asking for a written list of all assumptions made by them to arrive at their revised estimate. Again, a double check is preferable and it is sometimes possible to compare with previous projects, although this must be done with caution as no two systems are the same.

In order to make an accurate estimate of manpower, we need to take account of working days before any allowance is made for holidays, training, sickness etc. These allowances will vary from company to company, and country to country and will influence the elapsed time necessary for completion of the project. Figure 4.1 shows a typical breakdown of items which will be used to convert man days to elapsed days available for work. Note that Saturday and Sunday are always excluded – although they are often used for contingency towards the end of the project! An alternative approach is to assume that a working

Available working days per annum = 52 × 5 = 260

Public holidays	10	
Annual holidays	20	
Sickness	5	
Training	10	
(includes formal courses plus non-productive time such as company meetings)		
TOTAL	45	260

Net days available to work on = 215

Therefore:

$$\text{Elapsed days} = \text{Man days} \times \frac{260}{215}$$

$$= \text{Man days} \times 1.21$$

Fig. 4.1 Man days to elapsed days conversion – England

week is four man days and not five when converting from mantime to elapsed time. If the estimating is being done in terms of man hours, then it should always exclude overtime unless there are specific reasons for including it in the initial phases of the project. For example, in a development environment which is being shared by many projects, it is often impossible to do performance testing during the normal working week without adversely affecting all the other projects. Hence it may be necessary to schedule weekend working for the latter stages of acceptance testing to get sole use of the computer. In the overall implementation plan, including the user's commitments, it may again be necessary to include weekend working if the department concerned is constrained by resources such that these are all needed Monday to Friday to run existing systems and can be available only at weekends to assist in cut-over to the new system. If such extra hours are scheduled into the estimates, the additional costs must be accounted for; often weekend work is paid at $1\frac{1}{2}$ to 2 times prime shift rate, and the effort must be scheduled for as short a time as possible. Working seven days a week for anything more than four weeks rapidly reduces the effectiveness of all staff, be they development or user team members.

While detailed estimates follow, it is often useful to have a rough guideline to indicate the percentage of effort to be applied to the different phases of a project. Because of the diversity of project type, this must be treated as a very broadbrush appraisal to get the estimating exercise under way. It must also be remembered that the balance of effort will shift as the techniques for systems development migrate from the conventional to those using specific methodologies such as fourth-generation languages, databases and so on. Despite these variances, the project leader might like to use the following breakdown:

Feasibility	10%
Analysis inc. statement of requirements	15%
Design inc. system and program, logical and physical	20%
Programming and testing	40%
Installation, training, post-installation support	15%
	100%

4.2.2 Estimating feasibility studies

Because much of the work in a feasibility study is investigative, involving many interviews and much assimilation of data, the only way to estimate the effort and time-scales is to work on experience. If a number of previous projects have been completed in the organization, there is a chance that details will be available of the amount of time spent interviewing, assessing information from the interview and so on. This data should be used with care, but it will form a useful basis for the project leader. In the absence of such historical data, the effort shown in Figure 4.2 can be taken as typical of a medium-sized commercial project. For simple projects this may be an over-estimate and for complex real time systems the times will need to be extended, especially that taken to read and assimilate all relevant material. The project leader should encourage all his team members to maintain records of the effort expended on each of the tasks so that they are able to analyse them at the end of the project and update those statistics which are personal to their performance. Needless to say, he should also update his own personal statistics.

4.2.3 Estimating analysis and design phases

The pre-requisite to estimating the analysis and design phases of a project is a good plan with every task itemized and described in as much detail as possible. This exercise alone can take significant time if it is to be done thoroughly. The problem arises when the project requires to be

Activity	Effort
Interviewing	$\frac{1}{2}$ man day plus any travelling time per interview
Write-up of interview and secondary check of information	I man day per interview
Reading and assimilating existing information and results of interviews	I man day per 50 pages of A4
Report writing – this is a very personal statistic which each individual should maintain	10–20 typed pages per man day
Typing	20–30 A4 pages per typist per day
Corrections to Ist draft	15 mins/page to include Ist reading, correction and 2nd reading
Printing and binding	Varies extensively whether job done in-house or externally

Fig. 4.2 Effort chart – feasibility study

fully estimated accurately prior to final authorization. The danger is that the estimating will be skimped and broadbrush estimates given which may prove to be totally wrong when the project has been authorized and resources made available to do the detailed investigation. It is well worthwhile for the potential project leader to fight hard to get time and staff to do a thorough estimating exercise before commitment: it will benefit him personally and help the development team and user management. A starting point for estimating the two phases is shown in Figure 4.3, but it should be stressed that wherever possible the quoted figures should be adjusted to reflect the organization in which the project is being developed. Large, formal organizations will inevitably require a longer development time for projects simply because of the structure of decision-making and authorization. Very complicated real time systems will require longer discussions of the proposed system and larger specification of input/output, files and processing.

4.2.4 Estimating implementation

Using conventional analysis and design techniques followed by a conventional language leads to the popular method of estimating programming effort by using tried and tested coding rates against number of lines of code per module. Experience has shown that there is a relationship between the size of a program, measured in source statements, and the time it takes to write and deliver it. The relationship appears to be constant irrespective of the language, excluding fourth-generation languages. Thus, if the number of statements can be assessed accurately, using known coding rates per day, the number of days' effort can be calculated. However, life is rarely as simple as that and in practice the relationship is not linear; very small programs are proportionally quicker to write and very large ones take proportionally longer. The complexity of the programs also has a significant effect on the time to produce in relation to the mean coding rate, hence simple application programs are much easier to produce than basic system software programs of the same size. The method depends inherently on the skill of the 'statement counter' and the use of average coding rates. The project leader must be aware of these factors and try to establish an independent check on the 'statement counter', perhaps by getting him to estimate an existing program and compare his estimates with the actual effort expended to produce it. As for the project team, it is advisable to calibrate them in terms of coding rate against the mean value used in estimating the development effort.

The complexity of a module or program can be expressed in simple linear terms – say from 1 to 9, where 1 is very easy and 9 is extremely difficult. This scale can then be reflected in terms of coding rate per day, again based on the average programmer. So, for example, the project leader may decide that the following coding rate complexity applies:

Complexity factor	Statements per man day
1	25
2	22
3	20
4	18
5	15
6	12
7	10
8	7
9	5

If it is found difficult to allocate such a fine grid to the complexity of a program, a range of complexity from 1 to 5 should be used. Hopefully, statistics will be available from previous projects to give a realistic coding rate pertinent to the development environment, say:

Complexity factor	Statements per man day
1	25
2	20
3	15
4	10
5	5

Applying the above factors to a typical commercial accountancy system, Figure 4.4 gives an example of the use of this method of estimating.

Great care must be taken by the project leader to define precisely what is included in the rate of statements per day. Is documentation included, is module testing included, is link test included, and so on? If this is not clearly stated there is danger of double-counting effort or omitting it completely.

Assuming that the following aspects are allowed for in the development of each module – design, specification, flowcharting,

Activity	Estimate
Assimilate existing material	1 man day per 50 pages of A4 per team member
Initial fact-finding interviews with senior and knowledgeable members of user department	$\frac{1}{2}$ man day/interview with 3 interviews per interviewee
Documentary interviews, and assimilating the information into a coherent picture plus clarification with user	$\frac{1}{2}$ man day/interview + 1 man day/50 pages + $\frac{1}{2}$ man day/interview/team member
Prepare and circulate a detailed questionnaire to other user staff, plus a schedule of interviews with them	$\frac{1}{2}$ man day/interview
Do the user interviews above	$1\frac{1}{2}$ man days/interviewee + 1 man day/50 pages assimilation
Internal team discussion following interviews	1 man day/interviewee/team member
Document all interviews	1 man day/interviewee/team member

Fig. 4.3 Analysis and design phase estimates

Confirmation of detailed interview contents with user management	I man day per head of management
Agree and document fully existing system	10 man days/team member
Present to user management and interested parties	I man day/team member
Finalize any queries and get user management sign-off	I man day/team member
Produce objectives for new system and discuss fully with team	4 man days/team member
Define functional design and get user agreement	4 man days/team member
*A Define output screens and/or print layouts and discuss with users	I man day/team member/screen or print
Define user tasks and detail associated transactions	2 man days per task
Define contents of all files and list all inputs	$\frac{1}{2}$ man day/input + I man day/file
Prepare preliminary process descriptions	2 man days/process
Plan recovery and control requirements	I man day/team member
Document fully each process and calculate path length per transaction	3 man day/process
Calculate system performance	2 man day/team member
Repeat from *A until both project team and user satisfied	
Get formal user management sign-off	2 man days – project leader
Document fully design specification	15 man days/team member + typing and binding

Fig. 4.3 – (contd.)

Sub-system	Program	Title	Statements	Complexity	Man days
1	A	Batchpoint	700	1	28
	B	Edit customer file	1500	2	75
	C	Edit general ledger	1200	2	60
2	A	Transaction sort/merge	500	1	20
	B	Customer update	2300	3	153
	C	Control sort/merge	200	1	8
3	A	General ledger merge	600	1	24
	B	General ledger master match	1400	2	70
	C	General ledger update	1700	4	170
4	A	Print statements	300	1	12
	B	Print ledger	300	1	12
		Total		Total	
		Statements	10700	Man Days	~~479~~ 632

Thus using Figure 4.1, 479 man days = 560 elapsed days.

Fig. 4.4 Estimating implementation example

coding, module testing, link testing, documentation – Figure 4.5 shows typical percentages for elapsed time for each activity, both for applications programs and system programs.

The program size, complexity and coding rates technique for estimating software is probably the easiest and most widely used. Given a competent person to estimate the module sizes and reasonable historic data on coding rates, the method produces estimates to within $\pm 5\%$ of actual. This could be better but compares favourably with the comments reported in one of the multitude of computer periodicals – 'if we achieve within 150% of our estimates we are doing well'!

Another technique used for estimating program production in the implementation phase is the use of formulae for effort and elapsed times based on the number of data types to be processed (T), the number of files (F) and the number of procedures to be executed (S). These factors can all be extracted from a properly defined program specification. As a rule, the value of T will equal the number of record types described in the program specification. One unit of S should be allocated to each record type that requires an entirely different set of conditions to be

Activity	Applications programs (%)	System programs (%)
Design	10	15
Specification	10	7·5
Flowchart	15	10
Code	20	20
Module test	25	37·5
Link test	15	5
Documentation	5	5

Fig. 4.5 Breakdown by activity

considered in the program; for each file; and again for each separate action to be done in the program, e.g. process record type A, form batch print file etc.

Figure 4.6 details the construction of the formula so that the project leader can adjust the individual factors if he has first-hand experience of developing systems or access to historical data. Using the formula stated, we can see that a simple listing program with four record types ($T = 4$), two files ($F = 2$) and fourteen separate actions ($S = 20$) will take

$$\frac{(3 \times 4) + (4 \times 20)}{2} + \frac{(12 \times 4)}{2} + 28, \text{ i.e. 98 man hours.}$$

4.2.5 More advanced estimating techniques

4.2.5.1 *Statistical methods*

The methods described in the previous sections of this chapter can be applied to small-to-medium systems of varying complexity but once the projects become large with difficult applications, more sophisticated techniques will need to be used. Based on the knowledge that software estimating is an art, not a science, recourse must be made to statistical methods to allow for the uncertainty of software size and structure at different phases of the project, and to allow for man's innate nature of not wishing to be specific when put on the spot about software sizing.

Activity	Effort (hrs)	Assumptions
Read-in	14	2 days/program
Flowcharting	$\dfrac{T+S}{8}$	$\frac{1}{8}$ hr for each T & S
Block diagram	$\dfrac{T+S}{4}$	$\frac{1}{4}$ hr for each T & S
Coding	$\dfrac{T+S}{2}$	$\frac{1}{2}$ hr for each T & S
Test data preparation	$\dfrac{T+S}{4}$	Allow $\frac{1}{4}$ hr to prepare test data for each T & S; assumes worst case of one procedure test per test data type
Desk checking	$\dfrac{T+S}{8}$	
Test control	7	1 day to set up testing
Testing	$4\left[\dfrac{3T}{F}+\dfrac{S}{8}\right]$	4 hrs/test, 3 tests per T with 1 result per F. 8×5 per test. Assumes only 1 test can be done/day
Documentation	$\dfrac{T+S}{4}$	$\frac{1}{4}$ hr for each T & S
Review	7	1 day to review/programme

Therefore:

$$\text{Effort (hours)} = \frac{3T+4S}{2} + \frac{12T}{F} + 28$$

Fig. 4.6 Effort estimation by formula/program

The project leaders' response to a question like 'how big do you think the system will be?' generates an answer 'somewhere between 80 000 and 200 000 source statements'. A very wide range, but he knows he should be safe within those bounds. Remember, he knows little detail at this stage and his reply will be based on earlier experiences of similar systems. Using this estimate, we can calculate that the average, or expected, size will be $(80\,000 + 200\,000)/2 = 140\,000$ with a standard deviation of $(200\,000 - 80\,000)/6 = 20\,000$.

So the expected size of the software is $140\,000 \pm 20\,000$ statements, which implies that there is a 68% chance of the software being in the range 120 000 to 160 000 and a 99% chance it will be in the range 200 000 to 80 000 statements. This may seem a very wide range of size but, at the start of a project, very little is known and it is good for the project leader to realize this. As the project progresses, more knowledge accrues and the work can be broken down into more discrete units. Each unit can then be treated as for the whole project with maximum and minimum and standard deviations being calculated. By this stage there will be a number of people working on the project and it would be wise for the project leader to get them all together to form a group concensus of the size of each unit. There are, or course, drawbacks with the group concensus being influenced by strong or weak members and the results influenced by politics, but the Delphi technique can be utilized to overcome this. With the Delphi technique,

(a) The project leader gives each team member a specification form on which the estimate is recorded.
(b) Team members complete the form anonymously – they can ask questions of the project leader but not discuss the estimates with anyone else.
(c) The project leader summarizes the responses on the form or requests another iteration and the rationale behind it.
(d) The team members again complete the forms anonymously and the process is continued.
(e) No group discussion takes place throughout the process.

Figure 4.7 shows a typical form used for Delphi iterations. The inputs can be used to give maximum and minimum software size and standard deviation. Once the system design phase is in

Project Name: Date:

Project No:

Module Name:

The following estimates were got from iteration number x

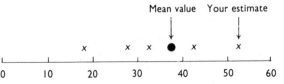

What is your estimate for iteration $x + 1$?

Comments on estimate:

Fig. 4.7 Delphi iteration form

progress, it is quite normal to expect the standard deviation to be less than half that at the outset and the expected size within one standard deviation of the original. In the example, this means that the software size could be 155 000 $\pm$ 9000 statements, which implies an uncertainty of approximately 12%. As design and module specification progresses, this uncertainty can be reduced still further.

It must be stressed that throughout the use of any estimating formulae, the variables are always subject to uncertainty to a greater or lesser extent. This leads the project leader into considering the risk associated with each of these uncertainties. There are a number of well-established techniques for risk analysis that can be applied to each of the key variables. Most involve the use of computers to simulate the effect of combining different values for each of the variables.

4.2.5.2 *The Constructive Cost Model (COCOMO) of Boehm*

Probably the most widely read book on software estimating is *Software Engineering Economics* by Barry W. Boehm. This devotes itself to a study of the software life-cycle and the development of cost models of increasing complexity to allow for all the variables affecting software development costs. These cost models have been derived from data collected on a large sample of projects and from theoretical work done by Boehm.

At the highest level, basic COCOMO is a simple formula for estimating the cost of a project solely as a function of its size in delivered source lines of code. Intermediate COCOMO estimates the cost of the project in terms of its size in delivered source lines, staff experience levels, computer hardware constraints and the use of modern programming techniques. The most accurate and detailed level of costing is provided by detailed COCOMO. This uses the 'cost driver' such as the level of staff skills to determine the software product cost by individual phase, sub-system and module.

The intermediate COCOMO model is more suitable for the cost estimation of the detailed stages of software development. It contains an additional fifteen variables used to predict the cost over and above the basic COCOMO variable of delivered source instructions.

The most accurate estimates can be made from the next level of cost model – the detailed COCOMO system. Needless to say, this needs the greatest time to develop and much more information about the system being produced. The total system is broken down to a three-level software hierarchy. The lowest level is the 'module level' and is described by the number of delivered source instructions and those cost drivers which tend to vary, such as complexity, program capability and the machine being used. Next is the 'sub-system level' described by the remainder of the cost drivers such as time, or software tools, which are relatively constant in modules but vary across sub-systems. The highest hierarchy is 'system level' which applies to relations of the overall project such as nominal effort and schedule equations. In addition to the hierarchy model, the detailed COCOMO method recognizes that differing stages of the project will encounter differing degrees of complexity, require different staff skills and so on. Using the above basic approaches, Boehm has developed estimating forms, procedures and equations to assist the project leader in estimating his man effort to be expended at all stages of the project. The use of these is a non-trivial task and time must be built into the original plan to do the work, and subsequently to update it as more accurate information becomes available. However, the mantime cost of software development is the major factor in the overall development budget and will continue to be so, thus the project leader must pay serious attention to defining it initially and refining it frequently throughout the development cycle.

As with planning and controlling projects, there are a number of automated systems available for estimating software costs. The project leader should evaluate these in the light of the project type and the development environment and ensure he does not spend too much – or too little – effort addressing this aspect of the project start-up. These tools obviously cost money; to buy the software and perhaps specific hardware, as well as money to pay for effort to set them up. The project leader should use his judgement to evaluate the cost-effectiveness of using such tools and techniques.

4.3 Estimating computer resource requirements

If anything, the task of estimating computer resources for development and live running is more difficult than estimating man effort. Much more attention has been paid to developing techniques and tools to assist the project leader with mantime estimating than has been directed at machine resource estimating. With the ever-increasing power per pound of modern hardware compared to the ever-increasing cost per man, this is probably a reasonable decision, but it does not help the computer resources manager who is trying to plan his machine requirements and operating schedules.

The techniques for estimating some system resources, such as disk storage, are quite straightforward and for others, such as central processor time, quite difficult. For disc space, the maximum file sizes should be calculated and a contingency added to allow for work file and temporary transaction files. During system development, it is usually possible to operate with sample files much smaller than the live file sizes. When estimating live files, always ensure that a sufficient allowance for data growth has been made and, if possible, get a prediction of the growth over the five years from acceptance date and allow for this – especially if you are being asked to propose a new hardware configuration.

The amount of processing time used will be a function of many variables, including the number of other users on the system. It will depend on the type of development – batch or on-line – the language used, the software development tools used, the proficiency of the program, whether or not documentation is entered directly on to the machine by the development team and so on. There are some simple rules of thumb which provide guidelines to the computer resources manager. For on-line development, do not assume that the developer will use more than 5 hours per day connect time per terminal. If the total man weeks for development of a conventional system is known, it is reasonable to assume that 50% of this effort will be spent at the terminal based on 25 hours/week. In a mixed development/live running environment, it has been found that 5% of terminal time is equal to the central processor time. Knowing the size of the development team, you can then work out the number of on-line terminals required.

Converting these resources to cost is straightforward if there is a set charge per connect hour and a price per cpu second. For example, assume there is a project team of four programmers developing a system over six months. The total man weeks is $4 \times 26 = 104$ and hence the terminal connect hours would be $25 \times 52 = 1300$ and the cpu hours 5% of 1300 or 65 hours.

Terminal connect time is usually charged in pounds per hour, and cpu seconds in pence.

Therefore, assuming a connect time of £8 per hour and 2.5 pence per cpu second,

$$\text{Terminal connect cost} = 1300 \times 8 \quad = £10\,400$$

$$\text{cpu cost} = \frac{65 \times 3600 \times 2.5}{100} \quad = £5\,850$$

$$\text{Total cost} = £16\,250$$

4.4 Estimating project overheads

Section 4.2 concentrated on estimating true effort needed for software development in man weeks. To this must be added all the necessary project overheads essential to the smooth running of a successful project. The word 'overhead' is often misinterpreted as meaning superfluous, but in the project context it applies to the many ancillary activities that form an integral part of the project. If a project has a team size of ten or more, it can justify a project secretary to maintain all records, schedule meetings and control and prepare documentation. If this person is not provided, then allowance must be made for the use of in-house word-processing facilities in terms of cost and turnaround times, especially when critical documents are being produced. Proof-reading of these documents by the team members must be allowed for at five minutes per A4 page for the first proof-reading and five minutes per page when checking corrections. If secretarial or clerical support is not available, an allowance of half an hour per day should be made from all team members as well as from the project leader to cover these activities.

A number of regular meetings will be held between the project leader and his team, at project set-up, project re-organizations, status reviews, company briefings. These should be accounted for as with a team of twelve a two-hour meeting can mean about a man week of effort absorbed. Similarly, regular meetings between the project leader and the user department must be accounted for – each one is unlikely to be less than two hours and if the user is remote from the development site, a half to one whole day could be taken up.

Throughout the project life, the project leader will be assessing his team's technical performance, commitment to the project and so on, but every six months he should do a formal review. If done thoroughly and constructively, this will take approximately half a day per team member, so again, with a team of twelve, this is greater than a man week's effort. The end of project reviews will take at least as long as more information on the individual's performance throughout the project will be discussed. There will also be a data-collection exercise at completion of the project which may well extend the final review session.

If a comprehensive project control system is in use, the effect of this must be allowed for because team members will have to spend time preparing work done to date for feeding into the project control system. Some user departments require the project leader to demonstrate the system at a few key milestones while it is being developed. This demonstration will not just happen, effort must be put into preparing and presenting it. For simple- to medium-complexity projects, a day or two should suffice for preparation and presentation. For complex real time civil and military projects, the demonstration activity is much longer and can run into months of elapsed time with much computer time being utilized. The project leader must plan for this and adjust his end target to accommodate it. Rehearsals must be held for which it may be necessary to co-opt other department staff from different projects. Special programs may have to be written just for the demonstration although the project leader must actively discourage this – he has enough real programs to produce!

Public and annual holidays are accounted for in the factor between mantime and elapsed time, but the project leader must schedule them in. If the holidays are lengthy, then the absence of one team member may involve rescheduling the whole team to ensure that milestones are reached. Time will have to be spent auditing the system and then, when it is ready for delivery, archiving will have to be done first and all paperwork prepared for hardware handover to the 'customer'.

4.5 Estimating tools

As with project planning and control, there are a number of proprietary techniques and software packages which are available to use as estimating tools. In Chapter 3 we identified SDS2 and PROMPT as two tools for planning; both of these have estimating routines as well. The SDS2 planning activity and control activities centre around the PERT technique and it is this which is used to produce estimates of mantime in a resource schedule. This schedule can be influenced by changing constraints, or adding new ones, to produce a smooth loading of staff on the project if it is required. The resource scheduler does not restrict itself to manpower; other resources can be estimated and produced provided the base data is entered by the project leader or his team members. Network analysis, commonly referred to as PERT, is the basis of many project control packages and resource planning is an integral part of all PERT systems. Many of these packages are now available on personal computers, and the project leader must survey the market to find the one most appropriate to the project and company needs. For any project of medium to high complexity, he should insist on having such a tool. The initial cost is often in the range of £5000 to £10 000 which, when compared with manpower costs, is insignificant. At current charge rates for professional staff, this would buy the project leader only 2–4 weeks of effort and yet the tool would be available over and over again. It will be used at frequent intervals throughout the project, not only at the monthly re-planning cycle as recommended earlier. It will allow the project leader to look at the effect of various courses of action dictated by decisions taken by himself or senior management. There is no doubt that for anything but the smallest projects, a computer-based estimating and planning tool is a valuable asset to the project leader.

4.6 Alternative estimating methods

In the earlier sections of this chapter, both simple and comprehensive estimating techniques have been described to evaluate the mantime resources required. Each of these has revolved around the reduction of effort from a knowledge of lines of code to be developed and the complexity of this code. Initial estimates are made with experience from previous work and these are refined to the specific project as time progresses and the number of cost constraints become clear. On the whole, these techniques have assumed that the time and resource is available to complete the job. In practice, this is seldom the case and the project leader often has to work within limited time, cost or manpower dictated either by the development department or the user. For example, a small commercial customer wants an order-processing system, it must fit on his existing 64 kb personal computer, he only has £10 000 to spend and it must be ready for use in three months. This situation does not allow the project leader much flexibility! This approach to estimating is not recommended as the software development is totally driven by external preconceived constraints which lead to an unprofessional approach to project planning and control and frequently an unprofessional end product.

In a systems house environment, there are many situations in which trade-offs must be negotiated to allow the manpower and machine estimates to reflect a cost which is going to win the business against the competition, and a 'price to win' must be computed. This approach can be very dangerous unless cost estimation is done very thoroughly and change controls are implemented in the final development to cover the areas of enhancement inevitably required by the customer. Computing a price to win will almost always mean a reduction in the delivered facilities against the ultimate requirements of the customer. Both this technique and a Parkinson estimate – where work expands to use all the resources available – are examples of top-down estimating, which rely on a reasonable degree of experience from previous projects and look at the proposed development at a high level. The danger is that by not examining the lower levels of sub-system and components, critical system or program logic is not identified, and this could have a major impact on the ultimate resources and price.

The converse of top-down estimating is bottom-up estimating, which is what has been described earlier in this chapter. Because proposed developments are looked at in such detail, the planning and estimating phases of the project start-up are much longer and the cost is usually higher than that arrived at from top-down estimating provided that system level costs have been thoroughly investigated. If the estimations have only looked at module costings, they will have omitted all the integration costs and overall system design and analysis.

Reference has already been made to the Delphi technique. This is a group meeting approach to estimating based on the experience – preferably of experts – of a group of people who are convened to a meeting to pool their experience anonymously! This anonymity ensures that a cross-section of views are applied to the problem without influences between group members; it eliminates the bias introduced by strong or weak personalities and interdepartmental politics, and gets to a consensus opinion.

4.7 Effect of design methodologies and languages

Design methodologies change the balance of effort in the various phases of development. Inevitably, much more time is devoted to the early stages of the project with close examination of data types and data flow. Hence the manpower requirements will be higher in analysis and design than in conventional systems development. As a result of this concentration of effort in the early stages of the project, the actual production of the code is reduced to a more mechanical operation requiring lower-level skills and is completed in a shorter time. There is then a shift in the manpower distribution curve for methodology-based projects. Use of a methodology will also increase the predictability of software estimating in that it will force similar projects to be broken down to similar units and hence comparison is much easier. The output from the decomposition of systems is consistent and bound to be documented so that baseline data for estimating is more visible, again helping with comparisons for subsequent projects.

The implication on machine resources is probably a higher peak in a shorter time during code development and testing. There is also an increasing requirement to provide computer time for tools used during the early phases of methodology-based projects. The precise balance of manpower and machine effort required will depend on the precise methodology proposed and the degree of automation included within that technique.

The major effect on estimating of different languages is the rate at which code can be produced. Thus the techniques for producing estimates will be little changed but the factors that convert lines of deliverable code to man weeks of effort will be drastically affected. Early systems were developed in basic machine languages and assemblers and, while the flexibility was excellent, the rate of coding was very slow and tedious. There is still a place for such languages in specialist routines within higher level language developments and where it is necessary to obtain high-speed processing within fixed storage restraints. High-level languages such as COBOL, CORAL, PASCAL and BASIC all make the task of writing programs much easier, and coding rates can be three to five times faster than for Assembler. There is a temptation to quote 'typical' lines of code per day but experience has shown that this is a figure very specific to a system, a user environment and a project team. Hence, use of such figures in estimating should be done only when historical data is available to give a reasonable confidence level to the information being used. Technology progress means that we are now dealing with fourth-generation languages where productivity can be ten times better than low-level assembler languages and, when used in conjunction with methodologies the actual coding phase of the project becomes extremely short. Not only short but very accurate: experience shows that module testing is frequently completed in one pass. Already, fifth-generation languages are being developed and promise even higher productivity, ease of use, speed and accuracy of testing. This means that project leaders and, more importantly, their managers must be prepared to wait for visible signs of code being generated until very late in the project, but when it comes it will come very quickly. Emphasis will have to be placed on building analysis and design skills to put at the disposal of the project leader.

4.8 Summary

The art of estimating has been examined in this chapter and guidelines presented for the project leader to use in his job of setting up and controlling a project. It is still an 'art', but much valuable experience is being accrued throughout the industry and more software products are appearing in the market to assist the newly-appointed project leader. These are too numerous to list in this book, but the project leader should make a thorough check of what suitable tools are available at the outset. There is a danger that he will be overwhelmed with suppliers' brochures and reference to an independent body for publications and advice is very desirable. Such information will be invaluable in helping to point the project leader in the right direction for a tool which could be applicable to his project. Many of the software estimating, planning and controlling tools becoming available will be based on personal computers which make them an attractive investment for all but the smallest project. A positive move by the project leader to adopt such tools in conjunction with development methodologies cannot fail to improve the quality and timeliness of software systems. This will in turn lead to more satisfied users.

5 Project control

5.1 Introduction

So far we have discussed the different phases in the system development process and identified the necessary planning processes for these phases, together with the estimation of resources needed. All the work of planning will, however, yield few benefits if the subsequent plans are not monitored and controlled effectively. Like planning, control is a continuing re-appraisal of the progress of the project against predetermined criteria throughout the development process. At a macro level, these criteria are time, quality and budget, but each breaks down to its own sets of targets. It is the monitoring of performance against these targets that forms the basis of this chapter.

As it is the project leader's responsibility to meet these milestone events, it is sensible to give him the option of choosing the specific control system he would like to use. He must be careful to choose a system that provides him with the correct level of control – too much detail means that too many resources are used just to control the project, yet too little detail means that insufficient data is available to enable effective control to be exercised. There may be constraints placed on the project leader if methodologies and project control tools are set as standards in the enterprise, but ideally each project should be assessed in its own right and simple tools used to control simple projects while reserving the resource-consuming tools for larger projects that can genuinely justify the expenditure of staff time and machine resources.

The project size and duration will often dictate the team's characteristics and this fact should be used to determine the formality of the reporting methods. Small, short projects imply that the project leader has a much more detailed day-to-day knowledge of progress than if the project requires many man years' effort and has an elapsed time of more that twelve months. In this latter environment, there should be sub-levels of control effected through team leaders through whom the project leader will exercise control. While a computer department should find the imposition of such controls commonplace, this is unlikely to be so with users, and some training may be necessary.

Any automated control mechanism, no matter how simple, must be operative from the beginning of the project, and the organization must exist to support the collection and entry of data, the processing of that data and its distribution to team members. On a very large project this could mean that three or four support staff are needed to run the project control system. The identification of the need for these resources lies with the project leader. This need will have to be supported by a sound justification which can be demonstrated to the project leader's manager, otherwise the project control group may be considered a mere overhead and efforts will be made to reduce or remove it. In these cases, the project leader must defend his position with examples from similar projects. It is all too easy for management unfamiliar with the development of complex systems to take the line 'you have a computer and a project control package so why do you need people as well?' Get the organization right at the start and the project will stand a much better chance of success. Typically a project with £1M software content would require a PERT-type computer-based package and a minimum of one project control officer. If the software content is higher and interlinked with the provision of a computer environment, development of special hardware and its linking to standard central processors so that development costs rise to perhaps £5M, there will be a need for three project control staff.

These examples assume the use of 'conventional' software packages which require the manual collection of progress data and distribution of output progress reports as well as setting-up and then updating the network. The move towards integrated project support environments (IPSEs) will change the emphasis from this manual support and may reduce the required numbers but at the expense of significantly more computer power. This is a trade-off which the project leader must assess and then convince his management of the desired course of action. At the end of the day, the purpose of controlling any project is to ensure that the development is cost-effective. If the work is being done by an internal department of the enterprise, this profitability will be reflected by completing on time, using the planned resources and then being able to release them for further work. If the project is being developed by a systems house, the profitability will be measured by a comparison of actual cost with the original estimate submitted to the customer.

5.2 Reporting structure

5.2.1 Project structures

There are two aspects to reporting structures: the project organization and the company organization, or supplier-to-client relationship within which the project organization must operate. Simple projects with teams of less than six staff at times of maximum activity will require a simple, uncomplicated structure where each team member is allocated tasks by, and reports back directly to, the project leader. The frequency of such feedback can be every two to three weeks and at every significant project milestone. The project leader will then have to report to his line manager and to the client manager, usually on a monthly basis; this is illustrated in Figure 5.1.

When the project size increases to between six and twenty team members, it is impossible for the project leader to control each individual personally so the team is broken down into logical units, each headed by a team leader. The units may reflect different sections of the application being developed or, for example, reflect the difference between systems programming and applications programming. The team functions will also differ during the phases of the project with analysis and design being completed by one team, implementation (coding and testing) by another, integration by a further team and, finally, acceptance by a team of computer department staff and user staff. While each team will have its own discrete responsibilities, it is essential to retain a thread of continuity in addition to the project leader throughout the development. The senior person in the analysis and

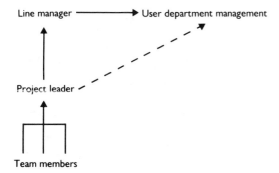

Fig. 5.1 Project reporting structure – simple product

design team, the leader for the implementation phase and a member of the design team could be allocated the development of the test suite, for example. Inevitably, there will be changes of staff throughout the life-cycle of the project and hence the project leader should structure his teams so that he can ensure continuity. A typical medium-sized project structure is shown in Figure 5.2, which reflects the phases of the development cycle, while Figure 5.3 illustrates a different structure system. It is up to the project leader to decide which structure best fits his particular situation. If he is unable to manage the project control tools personally, then a 'project control officer' should be appointed. In Figures 5.2 and 5.3, he is shown reporting directly to the project leader.

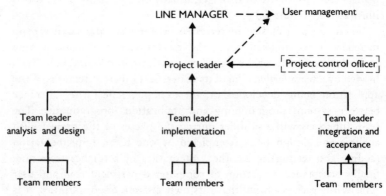

Fig. 5.2 Reporting structure – medium-size project

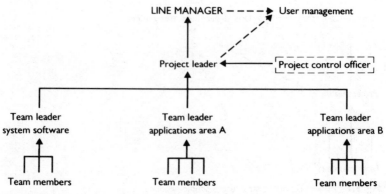

Fig. 5.3 Reporting structure – medium-size project

The structure in Figure 5.3 ensures a greater degree of continuity as the team leader is responsible for all phases of each application area, although the project leader must ensure that the interfaces between systems software and application areas A and B are concisely defined and adhered to during implementation.

If the team size extends beyond twenty and it is likely that the project time-scale is two or more years in duration, then a different team structure is required. Bearing in mind that every manager should allocate perhaps 20% of his time to each member of staff directly reporting to him, a complicated project will require intermediate levels of management between the project leader and the team leaders. Frequently projects of this size have a significant amount of hardware that needs configuring, controlling and maintaining as well as extensive system software and development tools. Figure 5.4 shows a typical large, complex project structure; such a project will always need a project

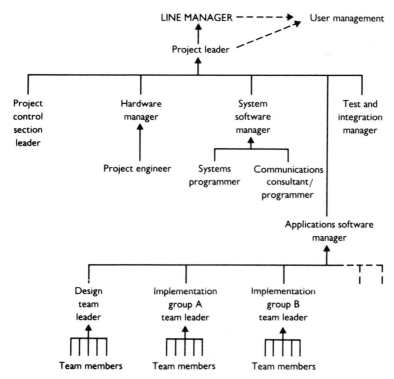

Fig. 5.4 Reporting structure – complicated project

control section to maintain progress, details from which the project leader can make planning decisions.

In all these projects, the need for software quality is paramount. In a simple project, quality assurance will usually be handled entirely within the project team by the project leader. As the development increases in size, while there must always be a quality assurance role in the project team, it becomes more important to have an external body to review quality independently and assess the technical correctness of the solution being implemented. In organizations which handle many projects, a permanent group of staff is often established to fulfil this role. In central government organizations, such as the Ministry of Defence in England and NASA in America where many very large projects are developed, whole departments are set up just to monitor the quality of work. Regular comprehensive inspections of work are carried out to ensure adherence to the quality assurance standards defined by the quality departments.

5.2.2 Company structure

In the last section we defined some typical project organizations based on size. These projects, however, also exist within enterprises which vary from small and simple to diverse and complex multi-national corporations. All types of project can exist in the large companies, but it is obviously unusual to find a very complicated project in a small one. In a small company, the data processing manager (DPM), if he exists, will usually report to a director, typically the financial director. This is illustrated in Figure 5.5 where the DPM is shown to have a responsibility to a business director as well as the financial director. This is not unusual as any system developed must serve the business objectives of the organization. Decisions on system development will usually be made jointly by these two directors.

In a medium-sized organization there may well be a board director responsible for data processing and management systems – the management services director. Reporting to him is either a data processing manager who controls all operational running and development of systems or a chief systems analyst and a number of development project leaders. Development decisions will be made by the management services director in consultation with the other board directors, and control of these projects will be by the management services director himself. This structure is shown in Figure 5.6, while Figure 5.7 extends the organization to that typical of a large

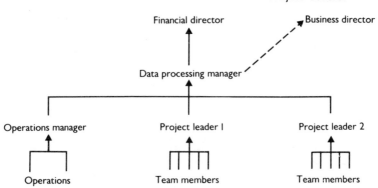

Fig. 5.5 Small company data-processing organization

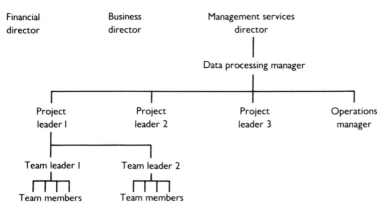

Fig. 5.6 Medium-size company data-processing organization

corporation. The development projects come under a systems development manager who reports to a management services director for line
functions but who is often responsible to a data processing steering
committee for new projects. The maintenance team sometimes comes
under the control of the operations manager and sometimes the systems
development manager.

All of these organizations represent companies that organize data
processing, and especially systems development activities, from their
own resources. A software service company or systems house deals with
development projects for many companies, often in diverse market
areas, and hence is organized differently. A typical structure is shown in

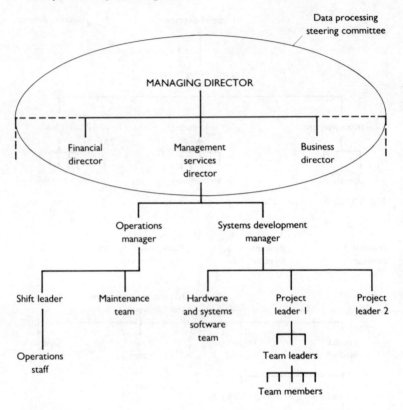

Fig. 5.7 Large corporation management services organization

Figure 5.8 where the responsibility for producing systems is placed with a production manager responsible for projects in one specific market area; production managers are then replicated for other market areas. To develop the wide range of applications required calls for the use of varied computers and staff with different backgrounds, and these resources are often co-ordinated by a resources manager. Systems houses also place great emphasis on the quality of the software produced and its ability to meet the customers' business requirements. Hence a quality manager will be found in the production organization responsible for the development of quality plans for all projects and the regular monitoring of such plans.

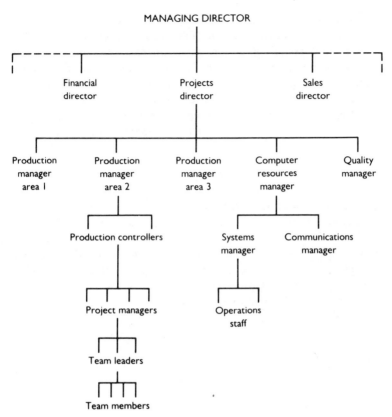

Fig. 5.8 Systems house organization

5.3 Reporting methods and techniques

5.3.1 Written reports

Reporting progress on a project can be done verbally or by written reports, informally or formally, frequently or infrequently. The choice of method lies with the project leader and should be suited to the type and size of project. However, there is one general rule: whenever there is discussion which results in a change to any part of the system, however small or large, it must be documented and signed off both by the development project leader and the user department representative.

The most satisfactory way of reporting progress on a project is to produce regular written reports. These should not be voluminous but concise and should concentrate firstly on recording the work done against a pre-determined plan and then on listing actions for completion against a specified date. There will, of course, be different types of report dependent upon the author and the recipient, but it is advisable to determine at the outset a standard format for each report and then retain it throughout the development cycle.

5.3.1.1 *Team leader to project leader report*

In a well-organized project, each team member will be given tasks of approximately two weeks duration. These will be reviewed by the team leader every two weeks and a report produced every two or four weeks to the project leader. It should summarize progress over the preceding period against the defined objectives, highlight problem areas, define corrective action and give the project leader the best estimate to completion for the team's modules. A typical layout would be:

Report identification
Review actions
Modules completed
Modules in progress
Modules not started
Estimates to complete (time and people)
Staff performance
Problem areas
Revised objectives for next period

5.3.1.2 *Project leader to line manager report*

This report should be comprehensive, detailing progress against major milestones, against the quality plan and giving not only progress vis-à-vis time-scales and budgets but also best estimates to completion. It also serves as a review of the team's performance to date and as a planning document, looking forward to activities still to be done and identifying the resources needed to complete them. In this way, re-planning can be combined with the periodic reporting of progress, so ensuring that re-planning is completed on a regular basis. The project leader should be honest when preparing his reports, especially with regard to time-scales and cost. Line management will be unable to assist if deviations from plan are not honestly reflected in the period report, but given sufficient warning it is often possible to take corrective action by diverting resources from different projects or re-negotiating deliverable dates. If the project leader does not notify his line manager until one week before completion that he is going to be four weeks late delivering the system, there is no chance to take corrective action and the project team suffers from a lack of credibility that seriously affects the user's view of the software and will not help the co-operation during the installation phase. So, be honest in reporting progress and highlight any slippages as early as possible. Figure 5.9 illustrates a typical period report from a services company where project control mechanisms are usually well developed. It illustrates very well the client relationships that are reported on, and could well be used by in-house development departments. Attached to this report will be the estimates to completion, updated every period, on the periodic budgetary control review form and the revised staffing levels on the planform.

5.3.1.3 *Project leader to user management report*

The salient factors regarding deliverables, time-scales and budgets should be given to the user management as well as to the line manager. This report, which is produced monthly, should however have a different bias to that produced for use within the development department. On the whole, the user is not interested in the problems, or solutions, of the development department. What he really wants to know is when his system is going to be ready for acceptance and what commitment he has to make, in money or resources, to reach this end. This report should therefore concentrate on the user's activities to effect implementation on time.

The layout of such a report should include:

Report identification
Narrative description of progress
Review of milestones – actual versus planned
Summary of user's resources to date
Detailed estimate of user resources to completion
Budget status
Change requests
User education programme
Implementation plan

Title	**PERIODIC REPORT FOR**		Page 1 of 4
			Sequence
Project		**Project No.**	**Date**

PROJECT MANAGER ....

PROJECT OFFICER .... PERIOD ....

ATTACH   (with changes marked in RED):
        PROJECT SCHEDULE, PLANFORMS, PROJECT BUDGETARY
        CONTROL REPORT

STATE PROGRESS AGAINST PLANNED TIMESCALES:

HAVE ALL OBJECTIVES/TASKS SET FOR PERIOD BEEN MET?   YES/NO
If NO, which have not, why and what is being done to recover slippage?

STATE DELIVERY DATES/MILESTONES
What is being done to recover slippage and is Client concerned?

As contracted	Last expected	New expected

HAVE STAFF RELEASE DATES CHANGED?   YES/NO
IF YES, which and why?

Fig. 5.9

| Title **PERIODIC REPORT FOR** | Page 2 of 4 |
| | Sequence |

LIST OBJECTIVES/TASKS FOR NEXT PERIOD

LIST WORST PROBLEM AREAS

LIST WORST PROBLEMS SOLVED (with explanation)

LIST POINTS REQUIRING DISCUSSION WITH CLIENT

LIST POINTS REQUIRING DISCUSSION WITH OTHERS (specify which)

LIST ANY CONTRACTUAL REQUIREMENTS WHICH MAY BE DIFFICULT TO MEET (e.g. delivery dates, specification requirements, performance criteria)

LIST ANY NEW PROBLEMS RELATING TO SPECIFICATION (omissions, error, ambiguities, inconsistencies, etc.)

Fig. 5.9 – (contd.)

Title		Page 3 of 4
	PERIODIC REPORT FOR	
		Sequence

	Attach details as appropriate
HAVE ANY SIGNIFICANT CHANGES BEEN MADE TO DESIGN THIS PERIOD?	YES/NO
HAVE ANY TECHNICAL MEMOS, WORKING PAPERS, ETC., BEEN ISSUED THIS PERIOD?	YES/NO
ANY NEW PROBLEMS RELATED TO HARDWARE OR CONFIGURATION?	YES/NO
ANY NEW PROBLEMS RELATED TO OPERATING SYSTEM, COMPILER, DBM OR OTHER THIRD PARTY SOFTWARE?	YES/NO
ANY UTILITIES, MACROS OR OTHER GENERALLY USEFUL SOFTWARE PRODUCED THIS PERIOD?	YES/NO
CHANGE CONTROL:	
HAVE ANY CHANGES BEEN REQUESTED BY CLIENT THIS PERIOD?	YES/NO
HAVE ANY CHANGES BEEN REQUESTED BY US THIS PERIOD?	YES/NO
HAVE ANY QUOTATIONS BEEN MADE THIS PERIOD?	YES/NO
ARE ANY QUOTATIONS REQUIRED?	YES/NO
ARE ANY QUOTATIONS AWAITING CLIENT DECISION?	YES/NO
DELIVERIES	
ANY DELIVERIES TO CLIENT THIS PERIOD (SOFTWARE, DOCUMENTATION, HARDWARE, ETC.)?	YES/NO
PROBLEMS DUE TO:	
STAFF PERFORMANCE?	YES/NO
TYPING AND ADMIN. SUPPORT?	YES/NO
QUALITY OF SOFTWARE?	YES/NO
IS THE PROJECT DOCUMENTATION UP-TO-DATE?	YES/NO

Fig. 5.9 – (contd.)

| Title **PERIODIC REPORT FOR** | **Page 4 of 4** |
| | **Sequence** |

IS QUALITY PLAN UP-TO-DATE AND BEING IMPLEMENTED?   YES/NO

WHAT IS NEXT QA ACTIVITY AND WHEN?
DO YOU REQUIRE OTHER SPECIALIST ADVICE   YES/NO

ARE THERE ANY OTHER ISSUES   YES/NO
(explain)

When is next PROGRESS MEETING with CLIENT?

          Date .... Location ....

SIGNATURE .... Date ....
          (Project Manager)

RETURN TO PROJECT ADMINISTRATOR

Fig. 5.9 – (contd.)

5.3.1.4 *Project leader to user committee report*

The contents of this report are similar to those listed in Section 5.3.1.3, but often serves as the basis for a meeting of individuals more familiar with the business than with computing. Care must be taken therefore to remove computer jargon and concentrate on the major issues of time-scales and budgets. The narrative description will need to be more comprehensive, and comprehensible, with less detail on future resources. The meeting itself is likely to be very formal, especially if the project is for a ministry department, local government, nationalized industry or large corporation, and hence the project leader should prepare himself thoroughly and be able to justify all the statements in the report. Associated with the meeting will be a set of minutes with actions which must be reviewed well in advance. Checks should be made to ensure that if the actions are not completed by comments in the report, separate memos are generated or the action is cleared by discussion at the meeting.

5.3.1.5 *Project leader to quality manager report*

In organizations which support an independent quality assurance function, it may be necessary to report on a regular basis to this function. Where software quality is handled within the project and within the line management function, the project leader's period report should cover all aspects relating to the quality plan. It may be that a copy of this report will suffice for the quality manager, but if not a separate document must be generated covering:

Report identification
Deliverables due this period
State of deliverables
Deviations from plan
Re-planned milestones and deliverables
Deviations from standards
Test plans status
Review of any remaining topics in the quality plan

5.3.2 Meetings

Regular progress reports serve only as a one-way communication medium and it is important to supplement them with regular progress meetings. At the beginning of the project, the project leader should plan a series of regular meetings to take place throughout the development process; preparation for these is important and time-consuming and this should be allowed for in the project plan. The level of documentation required to record decisions taken will depend on the meeting but, as a rule, concise statements and clearly-defined actions for named personnel taken by certain dates should be the goal. Meetings should be planned to achieve maximum results in mimimum time with the project leader as a tough chairman: time spent in meetings is time lost in developing software. To give of their best, team members need to feel involved in the total project and be able to identify the contribution they are personally making to its progress. This will be covered in more detail in Chapter 8, but it should be noted here that progress meetings can have a motivational impact on the team members. Members of this team in areas doing well will feel that they are getting visible praise and those in areas which are lagging behind will feel under an obligation to put in extra effort to improve their standing before the next general meeting. Hence, good preparation by the project leader can reap dividends while satisfying the information needs of the team.

5.3.2.1 *Programming team meetings*

Documented task lists with each task of approximately two weeks' duration will be given to team members on a continuing basis throughout the project. These tasks should be reviewed weekly, usually first thing on Monday mornings, with the team members. Progress should be noted, slippages re-planned, actions apportioned and workload adjusted between team members to allow the overall project to proceed at maximum speed. The meeting should be quick but effective: the use of a check sheet such as shown in Figure 5.10 will help to speed up the session.

5.3.2.2 *Project leader plus team leader(s) meeting*

A meeting of this type every two weeks should be frequent enough. The team leaders should distill the results of their weekly meetings to arrive at the impact on overall time-scales and budgets. By joint agreement, the balance of work should be adjusted between the teams to reduce overloads or take up slack, thus ensuring maximum productivity from the individuals. Project plans should be re-formulated based on the latest information available. The performance of individual team members should be discussed, recognition given to those working well and guidance plus job coaching or training given to those slipping behind schedule.

Project _____

Team _____ Progress meeting no. _____ Date _____

Task	Duration	Completion Date	Status (+ or − x days)	Revised Completion	Action by (Team Member)

Fig. 5.10 Progress meeting check sheet

The project leader should inform the team leaders of any change in contractual commitments, any system changes to be implemented under change control and any shift in attitude by the user department. Specific problems should be aired and, hopefully, solutions found. Actions should be placed on relevant individuals for completion by the next meeting.

5.3.2.3 *Project leader and user management meetings*

Like the reports written by the project leader for user management, these meetings should concentrate on the involvement of user resources and the impact these have on the project milestones. The performance of the user department in meeting deadlines for deliverables – such as test data – to the project team should be reviewed and forward plans adjusted accordingly. Any revisions to the requirement specification should be discussed so that the project leader will be able to estimate the time and cost of implementing these changes; changes already specified should be agreed by the user and formally signed off. Any problems likely to affect the project from within the development department or the user department should be discussed openly and avoidance plans formulated. Rather like period reporting, the best environment for the project is if both parties are open about their problems so that a common solution may be found. The meeting should be minuted and actions placed on individuals with time-scales.

These tend to be policy-setting meetings with the progress on the current project being placed in context with others in development or with activities being planned. More emphasis will be placed on the business effects implied by the installation of the system under development. The subsidiary effects of the system will be reviewed and appropriate actions identified.

Cost will feature highly in these meetings and the project leader should prepare costing sheets at various levels of detail to present to the committee and to use as back-up to detailed questioning. Overruns on cost should be examined by reference to the project ledgers or other accounting data. Any organizational changes forced by the system will be discussed, and the implications of the system for employment will be kept under review. Such a meeting can be hard work for the project leader but, if he is well prepared, it provides an ideal forum to present the project team in the best light and to push any ideas forward for discussion at the highest level. The meeting should be accurately minuted, with the minutes being circulated as quickly as possible.

5.4 Control systems and tools

In Chapter 3 on project planning, we identified a number of techniques and tools available for him during the planning phase of a project. As controlling a project is basically the comparison of 'actuals' against the planned values devised at the start, it is not surprising to find that the planning tools can be used by the project leader to do the controlling as well. He will, however, only be able to do the comparison of actuals against plan if he is measuring them, and forms for collecting such data will be described in Chapter 6.

5.4.1 Bar charts

Except for large projects, bar charts still remain the tool most likely to be used by project leaders to monitor progress. Taking a bar chart such as that illustrated in Figure 3.3 for a feasibility study, a vertical line is drawn at each time of review and the actual status of each activity is compared with the planned status. This is shown in Figure 5.11 where the project is being reviewed at the end of P5. By then activity B4, assessing the data, should have been completed but is actually one week behind schedule. Activity C1, defining the outline project plan, is precisely on schedule and activity D1, analysing savings, is three days ahead of schedule. As B4 is behind schedule, it has not been possible to progress as quickly as intended on activity D2, costing the system. Based on this information, the project leader is able to direct some effort from activity D1 to activity B4 at the start of period 6 which should enable

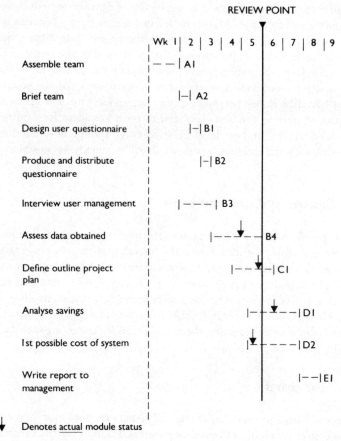

Fig. 5.11 Bar chart control

completion of this activity and consequently pull back on to schedule activity D2 by the next review at the end of P6. This is a simple example but illustrates how the project leader is easily able to prepare a visual picture of the state of his project. For more complicated projects he is able to produce subordinate bar charts covering various sections of the project which he is then able to review independently.

If computing resources are available and the size of the project warrants it, it is feasible to automate the bar chart reviewing procedure in connection with a formalized data-collection activity from the team members. One of the great benefits of doing this is the ability to easily retain historical data on coding rates, individual programmer productivity and the amount of code generated etc. This will be of significant value to subsequent projects.

With the proliferation of personal computers, the use of spreadsheets for planning has greatly increased. While not being strictly bar-charting tools, the end result when such techniques are used on project activities is not dissimilar. They allow the opportunity to update an original plan with actual performance and generate a revised plan with time-scales and costings. They are a valuable tool for the project leader, and selection of the most suitable package must be his decision. This in itself is not an easy task as there are a multitude of software packages available on a very wide range of hardware.

5.4.2 PERT-based control techniques

There are many proprietary systems available which rely upon the principles of network analysis, usually referred to as PERT, to provide the project leader with control information to run his project with. For simple projects, the manipulation of a PERT network can be done manually, but use of a computer-based product is highly recommended as in even the simple networks there is a lot of data manipulation. Like all control systems, the information provided from it is only as good as the data input. The project leader or project control officer must extract information from the team leaders and their staff regarding the progress on each activity identified in the network. As the system does not require the supplier of data to commit himself to an estimate to completion for the activity, there is a danger that the '95% complete' syndrome will prevail, and every attempt should therefore be made to get an accurate picture of how far the team have got through each activity. This data is fed into the network and the critical path is calculated. This is the shortest time to complete all activities, taking account of dependencies, and hence to complete the project. In a large project this will mean evaluating many alternative routes through the network which only a computer can do with accuracy and speed. All activities not on the critical path will have varying amounts of float, time that can be added to the duration for completion of these tasks without affecting the overall completion date of the project. The output from computerized PERT will give a listing of each activity and its float and will highlight those activities on the critical path. With this information, the project leader will be able to adjust his resources to improve the overall end date and utilize the floats. Some comprehensive PERT systems also aid in resource allocation by comparing actual critical path at the review date with planned completion date and adjusting resources accordingly. There is a wide range of implementations of network planning each with its own specific format and the project leader should investigate the applicability of each to his own project. He will undoubtedly gain benefits from using networks on all but the simplest of projects. The discipline requires a detailed definition of all tasks at the beginning and by looking for the critical path defines a time limit for the project. The measurement of achievement during the period under review is helped by looking at the impact on forward tasks and control. Resource distribution to best aid completion of the project is identified and will assist the project leader in allocation of tasks to his team.

5.4.3 PROMPT control system

We saw in Chapter 3 the contribution made to planning a project when using the PROMPT control system, currently recommended for use in UK government departments. The control aspects of this product are designed to compare actual performance with planned progress and hence detect problems well in advance, allowing re-planning to avoid any unmanageable situation. The objectives of such controls are to retain the business and technical integrity of the project. To this end there are two distinct sets of procedures, one to record actuals against planned at a series of reporting points and one to handle technical exceptions.

Remembering the management structure proposed by the PROMPT technique from project board to team leaders via stage manager, the control system must allow monitoring of progress at all levels. Hence there are four major control points:

- End-stage assessments
- Mid-stage assessments
- Quality assurance reviews
- Checkpoints

These are exercised primarily by formal meetings at which project status is reviewed and action plans formulated. The very first control imposed is that of a project initiation meeting, attended by the project board, stage manager and project assurance team, the activities at which are shown in Figures 5.12a and 5.12b.

Subsequent stage assessments review the status of the project and update the project plans, informing all interested parties formally of the results of these reviews. When an end-stage review has taken place, there is a strict sign-off procedure imposed to ensure that the project does not proceed to the next stage without being authorized.

Typical activities at end of stage review are shown in Figure 5.13. The mid-stage review can be covered by the project leader at any point during the development of the stage, or pre-planned into the overall schedule. The significance of an end of stage assessment is emphasized by the production of the mandatory 'acceptance letters' signed by members of the project board. These are the system acceptance letter,

Project initiation meeting

This meeting is called by the executive member of the project board in order to initiate the project formally. It is held at or near the commencement of the initiation stage.

The aims of the project initiation meeting

- To agree the project objectives
- To confirm project board and stage management responsibilities and levels of authority
- To decide the level of controls to be used throughout the project
- To approve the initiation stage plans
- To agree the initiation stage controls

Project initiation attendees

- Project board
- Stage manager (initiation stage)
- Project assurance team

Fig. 5.12a   Project initiation meeting

Project Initiation Procedure

Activity number	Description
P	*Preparation* (pre-meeting)
P1	Appoint project board and define their responsibilities
P2	Appoint initiation stage manager and define his responsibilities
P3	Appoint project assurance team and define their responsibilities
P4	Collect project initiation documents
P5	Determine project initiation meeting date
P6	Prepare initiation stage plans
P7	Identify initiation stage teams and their responsibilities
P8	Identify meeting location
P9	Inform attendees, distribute material
P10	Study stage plans and prepare meeting questions
R	*Review* (at the meeting)
R1	Review project management structure and individual responsibilities
R2	Present project objectives
R3	Present stage plans
R4	Approve initiation stage plans
R5	Define number of stages and control framework to be used throughout the project
R6	Confirm level of authority of project board members is commensurate with scope of the project
R7	Take meeting minutes
F	*Follow-up* (post-meeting)
F1	Review minutes and define any necessary follow-up action
F2	Document and distribute meeting minutes
F3	Authorize project initiation
or	
F4	Re-schedule meeting
F5	Report to IS strategy committee

Fig. 5.12b Project initiation procedure

user acceptance letter, operations acceptance letter and the business acceptance letter and are free format but must contain the following information:

- A summary of the project as seen by the author.
- Details of activities undertaken by the author to ensure the effectiveness of the work done to date.
- Identification of follow-up work to be done by the author or other managers.

The quality assurance review looks at the end product to be delivered at the end of the stage to determine its quality. Any errors or omissions must be documented and an action formulated to correct these deviations.

The full PROMPT control technique is substantial and comprehensive. It does require a significant amount of documentation, something which software developers are not very good at, but pays dividends in the timeliness and quality of the end system.

5.4.4 Software development systems (SDS)

Chapter 3 identified the need for more comprehensive planning and control tools to cope with large information system projects. One such tool from Software Sciences is the Software Development System (SDS); its planning aspects were described in Section 3.3.5 and its basic structure illustrated in Figure 3.6. The major strengths of this system are in its ability to control at a detailed level the progress of software development, documentation and configuration controls. This control is based on the system being able to store and manipulate implementation details, test plans and results, software versions, configurations, work plans, staffing, budgets and progress. For the central systems to be set up, it is necessary to de-compose the project down to its lowest level of activity and represent it as a hierarchy. Each activity must be estimated in time and resource: this resource can be commuted to cost by fixed charge rates for manpower and other facilities. A critical path analysis can then be done to establish the activity order with earliest and latest start and finish dates. As SDS2 is a database system, it is necessary to model the project as a set of entities to be held on the database. Each specific entity has associated with it records containing all its relevant details. All records relating to entities of the same type are held on a file, so that all activity records are held in the activity file and all person records in the person file. One strength of the SDS2 system is its ability

Activity number	Description
P	*Preparation* (pre-meeting)
P1	Determine assessment date
P2	Define assessment attendees
P3	Identify location of assessment
P4	Inform assessment attendees of date, time and location
P5	Update current stage plans
P6	Update project plans
P7	Prepare next stage plans
P8	Identify and appoint next stage teams and define their responsibilities
P9	Copy and distribute necessary papers to attendees
P10	Study papers and prepare meeting questions
R	*Review* (at the meeting)
R1	Present summary of stage and project status
R2	Review status
R3	Review off-specification list and planned actions
R4	Present next stage plans
R5	Approve next stage plans
R6	Take meeting minutes
F	*Follow-up* (post-meeting)
F1	Review minutes and define any necessary follow-up action
F2	Document and distribute meeting minutes
F3	Authorize progression to next stage
or	
F4	Reschedule assessment meeting
or	
F5	Recommend termination of the project
F6	Report to IS strategy committee

Fig. 5.13 Project stage review

to forge two-way links between records which are logically linked, e.g. BLOGGS works on ACTIV 1 so person record BLOGGS is linked to activity record ACTIV 1. Thus by defining the relationship between records, the software is able to represent a model of the entire project.

Input of the actual data is done on a periodic basis and a comparison made with planned data within the model. Sets of reports can be specified to the system for generation at the review points and, as it is possible to access the system via a visual display, the project leader can inspect the status of the project at any time and make changes to the database.

5.4.5 Integrated project support environments (IPSEs)

Surveys have shown that even the simplest support environment can yield benefits for the project leader by freeing him from the mundane, repetitive aspects of project control. Regular updating of work schedules, re-estimates to completion, re-calculating budgets from input on progress and cost rates, revised completion dates and resource allocation can all be done with the help of support software at far greater speed and accuracy than manually. For this reason the UK Alvey project is supporting the development of IPSEs. Many systems are planned to provide an experimentation facility which allows the project leader to try alternative resourcing and phasing of activities and to see the ensuing results before finally committing to one course of action. Such an exercise on anything but the simplest project would involve vast amounts of re-calculation which could not be entertained by the project leader without the use of an integrated project support environment.

Once project schedules have been accepted and verified by the project leader, these can be entered into the database of the support environment. Individual team member task lists are produced which can be output on hard copy or accessed via visual display units; regular reporting of progress against these tasks can be generated automatically.

Most IPSEs include a comprehensive resource scheduler, usually based on network planning so that the project leader will be able not only to experiment with different resourcing plans but will have the benefit of receiving regular consistent reports on project status from a confirmed basis of information. As individual project members will have limited access to the database for their own particular task lists, any errors in timing and workplans will be picked up by the individual and reported to the project leader. Corrections can then be applied to the database, schedules updated and reports generated.

Different IPSEs will support different design and implementation methodologies and support different languages. For example, a first-generation project support environment, 'Perspective', has been built for developing real time embedded systems in PASCAL using the MASCOT design methodology. The Alvey 'ECLIPSE' environment has been built to run initially on Digital Equipment VAX computers and will support ADA, PASCAL and 'C' programming languages with MASCOT and SSADM methodologies. The software tool 'MAESTRO' provides the facility to handle Jackson Structured Programming methodology (JSD) and Pseudo Design Language (PDL).

Over the next few years we shall see an increasing number of IPSEs appearing as commercial products. The project leader will then be faced with the same selection dilemma as he has today with PERT packages – to choose a software tool fitted to his project's needs. The golden rule must be to get a 'minimum fit' solution where there are enough facilities to provide adequate planning and control but with the minimum of initial expenditure and subsequent resource needs, both people and computers.

5.5 Impact of new developments

In a book such as this, it is impossible to define precise methods for managing all the vast range of project types and sizes which will be encountered in the development of new computer-based information systems. The tools selected by the project leader will have to be based on the specific project he has been allocated: simple tools for simple projects, comprehensive tools for complex projects is the general rule.

The implication of using comprehensive tools is increased computing resources to handle the processing load and data-storage requirements. With IPSEs, the aim is to retain all design, control and user documentation so as to ensure total system integrity. The implication of this for military command and control systems or for integrated business systems for large corporations and banking systems is the need for enormous amounts of disc storage, readily accessible by development teams. Because of the integrated nature of the control tools and the system itself, there is a need for stringent security controls to enable different levels of users to access only those parts they are permitted to see. Some users will have the ability to read and update parts of the system, others will only be able to read information. The project leader must give this serious thought at the start of the project to define the access levels for all development team members. In doing so, he will have to liaise with user management as they will be able to advise on the sensitivity of the programs and data to be handled. Security does not usually pose a problem provided it is carefully thought out at the start and individuals briefed fully as to their responsibilities. All access control levels must be set up in the support environment from the outset and controls applied to ensure that any violation is immediately reported to the project leader for his action.

Another implication of the use of IPSEs for software development is the need for additional training for team members in the use of such tools, and the project leader should build this into his budgets and time-scales. Most systems being developed offer the option of menu-driven access – which can be bypassed when the user feels competent – plus extensive 'help' facilities. This means that much training will be done while the project is in progress rather than through formal classroom sessions. This is probably better for the operators of the tools, but the project leader must realize it will affect productivity over the early phases.

Use of IPSEs and methodologies for development imply an increasing generation of data to be captured, processed, reported on and archived. In general, this will be handled on a centralized computing facility with access via visual display terminals, but an alternative or complementary approach has been the development of 'workbenches' for systems developers. These utilize the increasing power of the personal computer to concentrate design and programming activity at the individual's work position with recording of data locally. Most of these workbenches have very similar attributes: they are based on personal computers with integral hard discs; they are usually based around some form of data dictionary; they have intelligent editors which understand the graphics and syntax and automatically update the data dictionary as information is entered or changed. Each tool is usually geared to a specific methodology and provides some simple consistency checking.

'Excelerator' was developed in America and is marketed in the United Kingdom, being based around the Gane and Sarson method with James Martin data modelling. A central data dictionary supports a graphics editor geared to Gane and Sarson data flow diagrams, James Martin entity relationship data models and Yourdon-style structure charts. It permits the composition of sample screen layouts, reports and text with embedded graphic symbols for management reports.

'Speedbuilder' was developed by Michael Jackson Systems as a support tool for Jackson Structured Design (JSD) techniques. It too is centred around a data dictionary accessed via a context-sensitive editor prompting for information that needs to be recorded to document a JSD logical design. Structure diagrams to document model processes on to developed programs using JSP can be manipulated by linking to a program design facility. Similarly, text can be entered and maintained by linking to an editor. The product contains a report writer to allow the generation of customized reports from the data dictionary.

'Workbench I and II' is produced by Learmonth and Burchett Management Systems (LBMS) to support their LSDM methodology. The tool features a data dictionary supporting graphics and text editors to permit the intelligent entry of data flow diagrams, data models and entity life histories.

All the above products are available on the IBM PC running under PC-DOS and provide powerful tools at the fingertips of the systems developer for very low cost. A logical extension to their capability is to link them to the database of the IPSE held on a central machine. However, the project leader must exercise control over his undoubted excitement at having such powerful tools at his disposal and only utilize those which are strictly beneficial to the management of the project: 'tools for tools sake' will only waste time and money.

The use of methodologies and advanced generation languages remove the system developer further away from the code representation of systems. More and more tools are becoming available to develop and run systems, but usually at the expense of computing resources. However, with the power-to-cost ratio so attractive for hardware compared to labour costs, this is an attractive development.

5.6 Summary

This chapter has examined project control organizations and techniques applicable from the simplest project to advanced developments. Any new project leader is unlikely to be thrown into a large project controlled by an IPSE as his first assignment. Attention therefore to the techniques for simple projects will provide him with a sound basis on which to progress to more difficult jobs. His progression will usually involve him in an ever-increasing degree of automation in project control. The advent of the very powerful personal computer will do nothing but improve the accuracy of his control and reporting. While we have concentrated here on tools and techniques, the project leader must always pay just attention to the staff working on the development. The tools should be there to improve the quality of the system being developed and to help the staff achieve their goals. Tools which do this in a user-friendly way and which are innovative will motivate the teams; those which are cumbersome and only police the system will have an adverse effect. So, as project leader, evaluate carefully the automated assistance you give to your staff.

6 Project administration and review

6.1 Introduction

Documenting systems, producing reports and filling in timesheets seem to have a singular lack of appeal to computer analysts, designers and programmers; project administration is not a popular task for members of the systems project team. For the project leader, however, his skill in administration is important for the success of the project for it is only through close control, regular progress reviews and sound administration that he can be certain of delivering the new system on time and to budget. As with project control tools, the key is to establish just enough formal administration to manage the project and not so much as to burden down all the team members with unnecessary work.

As project leader, it is important to have a tight control on the team and while Chapter 8 will explain the management techniques available to do this, a sound administrative system will be of great benefit. With the advances in technology and increasing computer power, we saw in Chapter 5 some very powerful tools to help the system developers. How much of this power should be used for project administration is the project leader's decision. It will often be easier to get staff to key in information directly to a machine than to get them to report longhand. The most practical solution will probably be a compromise – some computerized and some manual.

Whatever administration system is set up, it is important to train the team in its use, start it at the beginning of the project and ensure that it is used regularly throughout its life. If a specific development methodology is being used, this will force the issue of documentation as progress will not be made through the project unless reports, reviews and minutes of meetings are produced at the correct stages. If, however, the project leader is to run a project without such a methodology, he will need to establish his own administrative system. There will obviously need to be a full set of files containing all documentation and all correspondence about the project; these should be clearly identified, split into logical divisions and always contain dated information. In addition, the project leader will need his own files containing data necessary to run the project. This should also be dated to ensure that there is a clear understanding of the sequence of events – this is especially pertinent if there is any dispute in the latter stages of the project about decisions made, or actions taken, in the early stages. Precise administrative procedures will depend upon the project leader, but in the rest of this chapter a number of items of general use are explained.

6.2 Project documentation

6.2.1 Private documents

There will be many 'public' documents produced during the life of a project, but the project leader will also need his own 'private' documents. A method of managing his own time to ensure he allocates sufficient to the relevant tasks within a project is essential. There are many proprietary products available which include a multitude of standard forms covering long-term planning, short-term planning, daily planners etc. It is not imperative that the project leader should adopt one of these but it is essential he controls his time to make every day an effective one. The basic principles of time management are explained in many publications and it should be the project leader's decision how to implement these.

There will be many diverse actions, conversations and decisions made during the development cycle of the project and too many for the project leader to remember. An extremely useful document for him to set up is the project diary: a standard large-page-a-day commercial diary is all that is required and the discipline to fill it in every day. Important facts should be recorded, notes on informal meetings, commitments to staff, agreements with users, discussions with other parts of the development organization. Salient facts should be recorded, not a lengthy narrative. If a commercial diary is not large enough, a standard A4 lined book could be used, dated the start of each new day. He should either carry the diary with him or ensure that it is updated at the end of every day; the information collected could be of use later in the project and will certainly help in the end at project review. This review will then be more comprehensive and complete and help subsequent project leaders in running their own projects.

6.2.2 Planning and control documents

Before getting to the stage of completed bar charts or network diagrams, many decisions will have been reached and many assumptions made; these should be written down so that the project leader is clear on the basis of the project plan and is able to refer back if needs be. These notes should be kept in the 'project control' file with the completed plans, and updated on a regular basis. Each version of bar chart and network diagram should contain a unique project reference, a version number if applicable and the issue date; before issue, they should be approved by the project leader and signed to this effect. Using standard forms, a front page can be set up with document title, author, date, issue number and authorizing authority. If this is kept on a word processor, updating is an extremely easy task.

In order to be able to apply change to these project plans, it is necessary to record progress and effort expended on a regular basis; this should be done weekly or two weekly by getting the team members to complete a timesheet showing the number of hours spent per day on each task allocated to the individual. Each task should be given a unique code consisting of the project name or code with a suffix for the task. If expenditure is incurred by the team member on behalf of the project, this should also be recorded against the same code. In order to encourage staff to complete these timesheets on the due date, it is a good idea to make recovery of expenses dependent on their submission, as shown in Figure 6.1! The timesheet should be used to record not only project tasks but holiday, sickness, training, attendance at conferences and so on, so that the project leader can see just how much time has been put into his project. This data is very useful for subsequent analysis at the end of the project.

For the team member to be able to report against tasks, he must have these clearly defined at the outset of his involvement with the project. The project leader will therefore have to document these in a workplan or task list. This must specify the project name or code, the activity to be undertaken, start and finish dates, and the relevant standards which apply to these tasks in terms of naming/coding conventions, use of

NAME		
STAFF No		

Time & Expenses Return

Period ending	

I declare that this is a true statement of hours worked and expenses incurred
SIGNED   DATE :

Project No.	PROJECT TITLE	TIME BOOKINGS (nearest ½ hour)																						Time Author-isation
		Mon	Tue	Wed	Thurs	Fri	Sat	Sun	Mon	Tue	Wed	Thurs	Fri	Sat	Sun	Mon	Tue	Wed	Thurs	Fri	Sat	Sun	Total Hours	
	TOTALS																							

EXPENSES SUMMARY

Project No.	Project Expenses					
	Accom Subsist & Travel	Small Consumable Items	Private Mileage Cost	Co Car Chargeable Miles	Author-isation	

NET TOTAL (ex VAT)			Total Miles
VAT Amount			
Gross Claimed			

Non-Project Expenses			
DESCRIPTION	AMOUNT	ACCOUNT CODE	Author-isation

NET TOTAL (ex VAT)	
VAT Amout	
Gross Claimed	

TOTAL EXPENSES CLAIMED	
VAT Amount	
Gross Claimed	

Overtime & Allowances				
PROJECT No	Chargeable Hours	Payable Hours	Chargeable Allowances	Payable Allowances
TOTALS				

Overtime & Allowance Authorisation
Expenses Authorisation

Fig. 6.1   Timesheet

common macros and so on; any dependencies or associated tasks should be noted, if only for information. A sample workplan is shown in Figure 6.2. An optional addition to this plan is enough space for the issuer to sign and the recipient to sign acceptance, although this is not usually necessary unless there are problems with commitment to the project by team members. If problems do exist, they should be tackled using the techniques described in Chapter 8.

While many projects will use existing hardware resources, there will be some which require the purchase of specific equipment or the use of external facilities such as bureau services. If the project leader is to be able to control his budgets, he will need to know the precise costs of such equipment or facilities and when that cost has been committed. This information should be supplied from the accounts department based on entries in their purchase ledger if external suppliers are used. If the services are provided by another department in the same organization an interdepartmental charging structure should be used. In both instances it will be necessary for the project leader to request the equipment or facilities, usually on a purchase requisition form stating

Workplan

Project name: Issued to:

Project no: Issued by:

                                                                    Date:

Task description	Task code	Man days	Start		Finish		Comments
			Planned	Actual	Planned	Actual	

Fig. 6.2 Workplan

details of the equipment or facility needed, the supplier, required delivery date and cost. He should find out the cost of such purchases before commitment to them and a copy of the purchase requisition should be kept in the project file. This can then be used as a checklist on the delivery of the items. It will not show when the enterprise actually pays the supplier but, from a project viewpoint, this is unimportant; the time of utilization is more significant.

We have discussed project plans, workplans and time recording earlier, but it is very helpful if the project leader has some documentation to use for summarizing this information and to enable him to compare actual results with predicted performance. This is particularly true for reviewing manpower usage, and the form illustrated in Figure 6.3 can be used for this as well as to record actual expenditure on expenses and additional facilities with a comparison against planned expenditure. Variations on this form can be developed to transfer man effort into cost and hence accumulate all cost information on one sheet. However, a budget summary form for manpower with subsistence expenses was shown in Figure 3.8. The project leader should decide the format he wants if he is not constrained by corporate standards. If he has a personal computer at his disposal, these forms comparing actual versus planned expenditure can be held easily on disc and the actuals updated on a monthly basis. Reduction to costs can then be easily achieved by specifying the daily cost rate per person.

We have seen in Chapter 4 that the effective working week is less than five man days because of the impact over the year of holidays and sickness. Sickness is unpredictable on the whole, but statutory holidays are fixed and annual holidays can be planned. Hence the project leader should allow for this when converting man effort into elapsed time for the project. It is important, therefore, that he knows when any absence from the team is intended, especially if staff are planning to have long vacations. The project leader is in the best position to see what effect such an absence would have on the development and therefore he should authorize all leave. A form to use for this is shown in Figure 6.4. If the project leader wishes, this form can also be used to show other non-project activities such as sickness, company meetings, training (career development) and so on. A sheet should be raised for each team member and kept with the project files. Requests for absence must be sent to the project leader and confirmed before such leave is taken. The further ahead such absence can be planned the better, as the project leader is able to build the information into his project plan.

PROJECT No. ....

			Effort B/F	\multicolumn{13}{c	}{PERIOD No. FINANCIAL YEAR:}				
NAME				1	2	3	4	5	6	7	8	9	10	11	12	TOTAL (C/F)
	Plan	days														
	Act															
	Plan	days														
	Act															
	Plan	days														
	Act															
	Plan	days														
	Act															
	Plan	days														
	Act															
	Plan	days														
	Act															
	Plan	days														
	Act															
	Plan	days														
	Act															
	Plan	days														
	Act															
	Plan	days														
	Act															
Periodic Labour Total – Plan																
Periodic Labour – Total – Act																
Expenses Comp. Facilities																
Bought Ledger Expenses (H/W)		£														
Bought Ledger Expenses (Other)		£														
T & E Expenses (H/W)		£														
Cumulative Expenses Total																
Periodic Grand Total		£														
Last Forecast Total		£														
Original Plan Total		£														

Title: MANPOWER & EXPENSES REVIEW

Page / Sequence / Date

Project / Project No. / Project Leader

DATE UPDATED ....

Fig. 6.3 Planned vs. actual resources

Project name

Team member

Annual leave entitlement:

Project start date:

Project end date:

Leave Requested (inc)		No. Days	No. Days o/s	Authorization	Sick Leave*	Stat. Hols*	Company Admin*	Non-Project Training*	Other*
From	To								

* Information to be entered from timesheets.

Fig. 6.4 Leave authorization/absence record

The need for formal written records to be kept of meetings and discussions has been mentioned frequently already. While this may seem excessive, readers who have been involved in a project where changes to requirements, designs or printed outputs have resulted from casual conversations with no supporting documents will realize the problems this can cause. It is particularly important to minute meetings accurately, apportion actions within those minutes and set time-scales for their completion. A standard format to the meetings and the minutes should be evolved with agreed distribution lists; a typical format is shown in Figure 6.5. If there are any contentious discussions and actions, then it is wise to check what has been written with all attendees before issuing the minutes. If the project is fortunate to have a project secretary able to take the minutes, then this task should always be allocated to him, otherwise at regular meetings it is wise to 'rotate' the taking of minutes and their distribution: it not only spreads the load but helps to remove any bias. Minutes should be issued promptly after the meeting has finished to allow people with actions to have maximum time to complete them.

Project:

Minutes of Meeting held on 'date' at 'time'

Present: List of attendees

1. Matters arising from previous minutes
2. Action review from previous minutes
3. Project status

	Actions	
	Name	Date

4. Any other business
5. Date of next meeting

Distribution: Attendees
+

Fig. 6.5 Minutes of meetings

6.2.3 Change control documents

The ability to specify the functional requirements of a system precisely depends on the size and complexity of the application. A simple payroll for a small company should be able to be defined concisely at the start, whereas the requirements for a weapons control system or an air traffic control system will be harder to define due to its complexity, the changing regulations during the development period and changes in technology directly affecting the end user. A mechanism is therefore needed to cope with changes to requirements during the progress of the project; it is aptly called change control. All changes need to be fully documented either by the user department or by the project team. An estimate must then be made of the effect of such changes on the overall system implementation. Modules which will be affected by the change will have to be identified and an estimate made of the change in resources implied by the revised design. This will in turn affect the project budget – usually increasing it – and end date of the project. It is up to the project leader to ensure that the user has accepted all these implications by getting him to sign the change request before making any revision to the existing project plan or scheduling of resources.

The amount of re-design and re-documentation is strictly related to the size of the change. Reduction in the capability of the end system may well not mean a reduction in the time-scale or cost of the project once implementation has started. The contents of a typical change request are shown in Figure 6.6.

Implementation of any change to the project content will have repercussions throughout the software and documentation. These changes must be recorded and all documents modified to be in step with the software. Equally, software version numbers must be updated and all associated test plans and test data kept in phase. In a project which is being controlled manually, the only way to ensure such consistency is to have attached to each change request a checklist of actions to be done before the change is signed off as being completely implemented and ready for acceptance. Such a checklist should ask the following questions:

● Is the change request authorized?
● Has the functional specification been updated?
● Has the project been re-planned?
● Are the design documents updated?
● Have the module specifications been updated?
● Has a revised test plan and test data been produced?
● Has the quality plan been revised?
● Is new acceptance test data ready?
● Has the handover documentation been updated?

It is in this area of version or configuration control that project control

Change Request: No. ....

Project Name: Change requested by ....

Project No.: Change estimated by ....
Date:

1. Narrative description of change and why needed.
2. Area of system affected (prime and dependent).
3. Modules to be re-designed and re-coded.
4. Effort required to implement change – mantime and machine time.
5. Cost of change.
6. Effect on project plan, existing milestones and planned end date.

7. Authorizations: (a) Re-estimates authorized by ....

Position ....

(b) Re-estimates accepted by ....

Position ....

(c) Authority to Proceed ....

Position ....

Note: The re-estimates are usually authorized by the project leader and accepted by his counterpart in the user organization. Authority to implement the change often has to go at least one level higher in the user management.

Fig. 6.6 Change request form

tools such as SDS2 and IPSEs are valuable as any change placed on a particular record in the database will automatically update all related records thus alleviating the danger of missing something in a manual system.

6.2.4 Staff appraisal documentation

So far we have concentrated on the control documentation associated with the project itself, but that relating to the performance of the project staff is also important. While a team is working together, there is a continuous assessment of the performance of individuals by the project leader (or team leaders). This assessment must be formalized by the project leader in the form of written project appraisals. For projects of around six month's duration, this can be done at the end, but for longer projects, some spreading over years, a review should be carried out every six months. It should concentrate on the technical knowledge acquired by the team member during the review period and how well he has applied it; his contribution to the team should also be reviewed. Areas of strength should be recognized and rewarded, areas of weakness should be addressed with action plans to improve them by training and supervised experience. Constructive criticism is valuable to the individual and the team.

Such project appraisals should form an input to an annual career appraisal to help plan the individual's future. Figures 6.7 and 6.8 are two examples of project and career appraisals which could be used for systems developers. A more detailed treatment of staff appraisal systems is given in Chapter 9.

6.3 Project reviews

6.3.1 Walkthroughs

It has been recognized that an improved software product can be produced if more than one team member contributes to the design or coding of a specific module. Traditionally, individual team members have been given modules to develop and there has been no interaction between these until the system integration phase and no separate view will have been taken of the efficiency of the coding apart from supervisory checks by the project leader. If, however, the programmer discusses his design and code with his peers, then new ideas will be generated that improve the software. Structured walkthroughs are an extension of this idea and an integral part of many design methodologies which are coming into use more and more.

6.3.2 In-project reviews

In Chapter 5 we identified the various control meetings necessary to ensure that the project leader is fully aware of the status of the project. These meetings will be held with the programming team and the design group to formulate a picture of the current state of the project. Internal project reviews are essential and must be documented in line with the recommendations in Chapter 5. For projects not using a formal development methodology, major reviews should be held at monthly intervals, although smaller projects will need these every two weeks. In both cases, action plans should be formulated and implemented. Typical reporting formats are shown in Figure 5.9 and can be adopted to suit the needs of the project leader. The various formalized project control approaches such as PROMPT force the issue of internal reviews by demanding them as part of the technique. Each review takes place at a specific stage of the project and is fully documented before it can proceed to the next stage. The extent of these reviews should be judged by the project leader in the light of the project size and his experience. There is always a danger, however, that because everyone is very closely involved with the details of the project, more wide-ranging issues may be overlooked: external reviews help to prevent this.

6.3.3 External reviews

It can be of great benefit to the project leader if an independent examination of the project can be made by a suitably qualified observer – perhaps by other project leaders. A short review, one to two days, can then be arranged where the project is viewed at a high level for design strategy, implementation strategy, testing techniques and solution of the business problem. Such a review should be very interactive between the project leaders, and the outcome should be well documented. It should be up to the project leader in charge to request such reviews, but one part-way through or at the end of each major phase can be useful. If no one else in the organization is qualified to execute such a review, the enterprise could employ a consultant from a reputable software house to do the job. This will take a little longer as the consultant will have to become familiar with the enterprise's operation in order to judge the project in its operational environment, but it does have the advantage of providing a truly independent review, devoid of internal company pressures. Also, if the consultant is carefully chosen he can bring many years of project development experience to bear on the problem. It is difficult to say how long these reviews should take but is unlikely to be less than a week or more than a month, unless a very complicated project is being assessed. Obviously a full report should be requested of the consultant, highlighting the recommendations to be implemented during the remainder of the project.

If the company is fortunate enough to have a technical auditing department, as many systems houses do, these should be used by the project leader at each major phase. Not only will the strategy be reviewed, but samples of the detailed implementation can be assessed

PROJECT APPRAISAL

NAME: .... GRADE: ....

PROJECT TITLE (if appropriate) ....

EVALUATION PERIOD – FROM:.. TO: ....

JOB LOCATION: ....

COMPLETED BY: .... POSITION: ....

READ INSTRUCTIONS BEFORE COMPLETING FORM

i   This report is to serve as a summary of information relating to the performance achievement and training/development needs of an individual. It should **always** be completed at the end of a project and **at least at six-monthly intervals.**

ii   It is important that the report is completed in a careful, honest and impartial manner. Individuals must be evaluated according to the standards of their present grade and in relation to their experience.

iii   **You must go through the completed form with the individual, who must sign it to indicate that he/she has read it.**

iv   Please note that Sections 7 and 8 should be completed during the appraisal interview and need a separate signature to authorise the training recommendation.

v   The original must be returned to Personnel Department for action on training and for inclusion in the person's file after it has been read and signed by the appraisor's manager.

1.   **APPRAISOR'S SUMMARY** (complete after Sections 4 – 8)

Signed .... Date ....

2.   **INDIVIDUAL'S COMMENTS ON THE APPRAISAL**

Signed .... Date ....

3.   **COMMENTS OF APPRAISOR'S MANAGER**

Signed .... Date ....

Fig. 6.7   Project appraisal

4. **PROJECT OR WORK SINCE LAST APPRAISAL:**
(Brief description details below)

Size of team .... Duration ....

Computer .... Language ....

Op. System ....

Individual's position   No. of staff
within team .... supervised ....

Time on project (months) ....

....

....

5. **INDIVIDUAL'S CONTRIBUTION:** (if job description printed, please attach)

ASSIGNED TASKS (In order of importance)	ACHIEVEMENT ASSESSMENT (how far were tasks successfully completed)	COMMENTS RESULTING FROM INTERVIEW

Fig. 6.7 – (contd.)

6. **PERFORMANCE ATTRIBUTES**

Listed below are selected aspects of performance which have a significant bearing upon the ability to work at higher level. Each aspect is described in terms of Outstanding Performance 'A' and Unsatisfactory Performance 'F'.

Rating 'A' or 'F' should be given if it is believed that one of the two statements is generally true and could be supported, if necessary, by specific occurrences. The intermediate ratings represent behaviour between these extremes.

		A	B	C	D	E	F		COMMENTS
Technical Performance	Highly effective in the practical application of professional and technical skills and knowledge							Does not apply professional and technical skills and knowledge to good effect	
Technical Knowledge	Very well equipped with the necessary depth and/or breadth of up-to-date technical knowledge							Displays serious gaps, weaknesses or limitations in technical knowledge required at this level	
Team Responsibility	Provides an excellent contribution as a team member or manager							Not a good team member or manager	
	Earns respect, cooperates well, organizes and motivates staff to give of their best (as appropriate to this level)							Does not earn respect, uncooperative and intolerant, inefficient in the use of staff, engenders low morale (as appropriate to his/her level)	
Client Relationship	Inspires the respect and confidence of all client representatives at the appropriate levels							Leaves client representatives unimpressed and lacking confidence in the company.	
	Communicates well orally and on paper							A poor communicator	
Commercial Awareness	Accepts full responsibility for time-scales and budgets and takes positive corrective action when necessary.							Ignores timescales and budgets	
	Spots and utilizes opportunities to secure business and commercial advantage for the company							Is not aware of business and commercial opportunities	
Drive and Penetration	Wholehearted application to tasks							Lacks energy	
	Gets straight to the root of a problem							Seldom sees below the surface of a problem	
	Determined to carry a task through to the end							Easily baulked by minor setbacks or opposition	
Company Contribution	Seeks and accepts responsibility							Reluctant to accept responsibility	
	Reliable							Unreliable	
	Prepared to put up with personal inconvenience or throw in extra effort when needed without complaining							Not prepared to put himself/herself out for the company	

Fig. 6.7 – (contd.)

7. **CAREER DEVELOPMENT:**
Summary of discussion of individual's career development, aspirations and plans.

8. **TRAINING NEEDS:**
Summary of discussion of individual's and appraisor's agreed recommendations for training.

Signed (appraisor) ....

Date ....

Please continue on separate sheet if required.

Fig. 6.7 – (contd.)

CAREER APPRAISAL

Surname .... Christian Name .... Grade .... Age Now ....

Date of appointment to grade .... Date of this appraisal .... Last Appraisal ....

Manager ....

PROCEDURE

1. Manager sends form to appraisee, who completes section 1 and left-hand side of Sections 2, 3, 4, 5, 6 and 7, and returns it to Manager.

2. After the appraisal interview the Manager completes right-hand side of Sections 2, 3, 4, 5, 6, and 7.

3. Manager completes Section 8, the individual adds his comments and the form is returned to the Personnel Department for retention.

4. Manager is responsible for the implementation of the action list.

1. **WHAT JOBS HAVE YOU WORKED ON SINCE LAST APPRAISAL?** (Use blank continuation sheets if necessary)

Project	Position in Team	Duration	Responsibilities	Name of Project Manager/Controller

2.   LIST NEW SKILLS ACQUIRED ON EACH OF THE   APPRAISOR'S COMMENTS
      FOREGOING JOBS

3.   ASSESS YOUR PERFORMANCE ON EACH OF THE
      FOLLOWING JOBS
      (INCLUDE CLIENT'S VIEW AS APPROPRIATE)

4.   LIST AREAS OF WORK IN WHICH YOU COULD NOT   APPRAISOR'S COMMENTS
      PERFORM TO YOUR SATISFACTION. GIVE REASONS

Fig. 6.8 – (contd.)

5. CRITICISE YOURSELF CONSTRUCTIVELY UNDER THE
   FOLLOWING:

   APPRAISOR'S COMMENTS

   a) **Planning**

   b) **Communication** (Face to face; Written; Presentations)

   c) **Management** (own; others)

   d) **Personal Qualities**

6. LIST AREAS OF TRAINING AND/OR PROJECT WORK
   YOU FEEL WOULD BENEFIT YOU IN THE FUTURE

   APPRAISOR'S COMMENTS

7. LIST ANY SPECIAL POINTS YOU WISH TO DISCUSS
   AT YOUR APPRAISAL

Fig. 6.8 – (contd.)

8. **APPRAISAL SUMMARY**

8.1 **PERFORMANCE ASSESSMENT** (to be completed by Appraisor)

	A	B	C	D	E	F
i) Technical performance						
ii) Technical knowledge						
iii) Team responsibility						
iv) Client relationship						
v) Commercial awareness						
vi) Drive/penetration						
vii) Company contribution						

8.2 **INDIVIDUAL'S COMMENT & ACTION LIST**

Signature .... Date ....

Fig. 6.8 – (contd.)

and advice given for the future stages of the project. Experience has shown that such reviews always generate new ideas and improvements to the development of projects. The quality plan should also be reviewed on a regular basis by the appropriate quality authority which may be the project leader's line manager in a small organization or a separate department in a large enterprise. The key factor is to get the plan reviewed at points established by the project leader in the project plan. For this review to be comprehensive, the project leader should be able to demonstrate his records of compliance with the quality system, project control system, design and implementation methodologies. He should be able to show that configuration control has been rigidly applied and that testing has been comprehensively executed.

6.3.4 End-of-project review

We have concentrated so far on ensuring that reviews are held throughout the duration of the development so that deadlines can be met, deliverables provided as defined in the quality plan and all controls efficiently implemented. Provided that this has all happened, the project leader will be able to feel satisfied when the system has finally been accepted by the user department. However, one final review remains to be held: staff appraisals should be completed before the development project is completed. As stringent controls have been exercised throughout the project, a large amount of data will have been collected of actual performance against planned progress, the usefulness of development tools, problems with suppliers of external services, team performance and so on. This information should now be collated and analysed to pass on to future projects so as to improve their performance even further. If this activity is consistently done at the end of every project, a substantial 'database' will be available to call upon for planning future ones. The review format shown in Figure 6.9 is typical of an end-of-project review although variations can be made to suit the specific environment in which the project was developed.

6.4 Summary

Administration and review techniques are important to the success of a project. Development must be carried out in an ordered environment with the careful recording of all appropriate information. This may seem tedious to start with but it is the only way to run a successful project which meets the user's business objectives. By taking such an approach, a 'professional' image is given to the project, and team members will learn many lessons which they will be able to apply when in the position of project leader themselves.

As time progresses we may well see the implementation of an integrated project support environment on a personal computer – this will give real power to the project leader of the future. Use of more formal methodologies on a wider basis and the progressive refinement of programming languages will lead to much more self-documentation of systems. This will reduce the administration load on the system developer and leave him free to concentrate on the design and implementation of the system itself. However, there will still need to be an effective project manager to control the use of advanced tools and deal with all the management aspects of a project team.

Title		Page 1 of 5
PROJECT REVIEW		
		Sequence
Project	**Project No.**	**Date**

PROJECT MANAGER ....

CONTRACT TYPE:

TYPE OF WORK:

PHASES OF WORK:

TYPE OF SYSTEM:

PROCESSOR SYSTEM: ....

DURATION: .... From .... To ....

TOTAL EFFORT: .... (man days)   MAX. No. STAFF ....

BRIEF DESCRIPTION OF APPLICATION:

DISTRIBUTION

       RETURN FORM AND ATTACHMENTS TO PROJECT
       ADMINISTRATOR

Fig. 6.9 Project review

158 Systems Project Management

Title	Page
PROJECT REVIEW	**2 of 5**

DEVELOPMENT ENVIRONMENT
   Computer(s) ....
   Operating System ....
   Compilers(s) ....
   Data Management ....
   Terminal Management ....
   Other (specify) ....
   ....
   ....
   ....

SUMMARISE EQUIPMENT CONFIGURATION:

TARGET ENVIRONMENT
   Computer(s) ....
   Operating System ....
   Compilers(s) ....
   Data Management ....
   Terminal Management ....
   Other (specify) ....
   ....
   ....

SUMMARISE EQUIPMENT CONFIGURATION:

Size of Total Software code
(k bytes) ....   Size of Data Areas .... k bytes
Number Programs/Tasks ....   Total Lines of Code ....
Memory Occupancy (k bytes) ....   Percentage New Code .... %
Total Memory Occupancy (k bytes)   Total Memory Availabile .... k bytes

....
(including manufacturer's S/W)

METHODS & TOOLS USED
   Analysis Method ....
   Design Method ....
   Programming Method ....
   Programming Tools ....
   Test Methods ....
   Test Tools ....
   QA Tools ....
   Documentation Tools ....
   Management Tools ....
   Other Tools ....

Fig. 6.9 – *(contd.)*

Title				Page	
		PROJECT REVIEW			3 of 5

TEAM

Name	Grade	Role(s)	Relevant previous experience	Man days worked

EFFORT DISTRIBUTION	Actual Man Days	%	Estimated Man Days	Comments
Staff Training	....	....	....	....
Management & Client Liaison	....	....	....	....
Feasibility Study	....	....	....	....
Requirements Analysis	....	....	....	....
Preliminary Design	....	....	....	....
Test Scripting/Acceptance Test Spec.	....	....	....	....
Sytem Design	....	....	....	....
Implementation	....	....	....	....
Integration & Test	....	....	....	....
Acceptance Testing	....	....	....	....
Project Engineering	....	....	....	....
....	....	....	....	....

TEAM PERFORMANCE
A Staff Project Appraisal form should be completed for each person. Comment below on how well the team fitted together: whether it contained the right mix of skills and how well the members worked together as a team; worst problems.

Fig. 6.9 – (contd.)

Title	Page
PROJECT REVIEW	4 of 5

PERFORMANCE
  Give reasons for deviations of greater than 10% between original estimate and actual for each of the above.

Outside Service

  Computing/Data Prep.
  If outside computer or data prep. or other hardware/services were used, comment on quality, cost, etc. Should we use them again?

  Sub-Contractors
  Comment on quality, cost, etc. Should we use them again?

  Client Relations
  Was the client satisfied with our work?

ASSESSMENT
  Technical Performance
  How well did the job go technically? List worst problems.

Fig. 6.9 – (contd.)

Title	Page
PROJECT REVIEW	5 of 5

PROJECT MANAGER

Lessons learnt – What lessons did we learn from the project?

Signed .... Date ....
               Project Manager

PRODUCTION CONTROLLER

Lessons learnt – What lessons did we learn from the project?

Signed .... Date ....
               Production Controller

Fig. 6.9 – (contd.)

7 Choosing the team

7.1 Introduction

Project managers don't always have the opportunity to choose all the members of their project teams. In many cases, the project manager is told who his team members will be; in other cases there is some element of choice from those available members of permanent staff who are not assigned to projects, and occasionally the project manager has the opportunity to build a team from scratch using permanent members of staff, contractors or people who have been specially recruited for the project. In this chapter, we have assumed that the project manager has a completely free hand to select and build his team; in this way, all the aspects of team selection can be covered.

The quality of the end system relies substantially on the quality of the people in the team. Project leaders need to take careful decisions about the types of people the team needs and the range of skills the project demands. In addition to their technical skills, for example, programmers must be capable of building effective working relationships with one another and, in many cases, with the client; the analyst and designer may well have to convert reluctant users and give support and guidance to programmers. It is not necessarily the case that first-rate programming and analytical skills go hand-in-hand with first-rate interpersonal skills. A project manager needs to prioritize these different skills and select a team with the optimum abilities to carry out his particular project, always mindful of the financial constraints placed on it. Although we talk of selecting a team, at this stage it is seeking individuals who will eventually form a work group which can finally be developed into a team. The abilities of the team as a whole should be greater than the sum of the abilities of each member, and throughout the recruitment and selection process it should be borne in mind that not only are we looking for individuals who shine as individuals but also people who can motivate themselves to work for the group goals.

7.2 Job and employee specifications

Before searching for candidates, it is essential for the project leader to be clear about what he is looking for: an ill-defined search is a waste of time and money. If he doesn't know what he wants, how will he know where to look for it, or know when he has found it?

First, the project leader needs to define the job and the tasks to be performed, the circumstances and the environment in which they will be performed and the nature of the required end result with the defined standards and quality to be achieved. A job specification or job description is therefore needed, covering what the individual has to do in his job, and how, where, with what and with whom. An example of a job description form is shown in Figure 7.1.

The information needed for an accurate and comprehensive job specification must be gathered systematically and recorded: it is needed for the present project and for future ones. Indeed, standard job specifications may be available, but if not the usual method is to set out separate ones for the different people to be employed so that, when they are put together, it means that all the tasks of the project are covered and its aims can be achieved. Each job specification needs to answer the following questions:

- What are the main duties of this job?
- Should special note be taken of any responsibilities the holder has for the work of others, either within the team, for the user, or for equipment, safety etc.?
- Which are the most common difficulties encountered in this job?
- Are working places noisy, isolated etc.? Will work be unvaried, under pressure (what sort of pressure?), away from home?
- Will the incumbent have companionship, work in a team or in isolation?
- What type of people does the job-holder contact; who initiates this contact?
- What amount or degree of supervision is to be received and given? Is there freedom of decision-taking?
- Are there irksome conditions?

The project leader can get help in the preparation of the job specification from people who have worked on similar projects or from a professional personnel department, if there is one. He can also read up information on past projects and evaluate where job specifications have

JOB DESCRIPTION FORM

JOB TITLE **DIVISION/LOCATION** **REPORTS TO** (JOB TITLE) **DIRECTLY SUPERVISES** (JOB/TITLE NO'S)	**LEVEL:** Draw a 3 level Organisation showing your JOB IN THE MIDDLE
OVERALL RESPONSIBILITY FOR STAFF NO'S	

COMPLETED BY: **DATE:**   **JOB TITLE:**

OVERALL PURPOSE OF THE JOB: (Why does it exist?)

CONTACTS: (In the course of a normal day, with whom does the job-holder liaise/influence eg. customers, other departments, senior members of the organisation)

SCOPE OF DECISIONS: (Financial, operational, at what point is higher authority sought?)

CAREER PATHS: (Promotional paths, other roles into which job holder may logically progress)

Fig. 7.1   Job description form

been adequately or inadequately designed. Recording job specification information from previous projects shows its value here.

An alternative approach might be to define the overall aims to be achieved by the project, list the tasks to be fulfilled by each specific job and create a skills and work experience audit. The project leaders could then recruit people one by one according to the remaining necessary skills and experience until he fulfils all the requirements of the audit. The main benefit of this approach is that it reduces considerably the time needed to recruit the first 50–70% of people. The selection problem, however, becomes more difficult when filling the final jobs and, moreover, a continual re-working of objectives is often necessary if this approach is used.

Job specifications are used to prepare the employee specifications that list the essential human attributes required to perform the tasks or deliver the outputs described in the job specification. The project leader needs to be careful not to over-specify essential requirements, making recruitment virtually impossible. At the same time, however, certain minimum requirements must be met in order for the project to be completed successfully. He must be clear about, and define, the minimum requirements which are acceptable for the project; additional requirements can then be defined in order of preference. An example of an employee specification form is shown in Figure 7.2.

By defining acceptable and desirable standards, the project leader can decide whether each job demands technical specialization, or all-round knowledge, or whether it can be done by a trainee; perhaps a certain job could be done by someone who is an expert in one field and will need to learn another computer language on-the-job, or by someone with all-round knowledge who may need training in a particular aspect of the job.

The project leader may decide to recruit someone with the minimum technical skills necessary as he or she is keen to accept responsibility for the job or keen to work in a team environment. He could consider adapting the computer skills of a potential team member to the demands of the project. Above all, if he is as flexible as possible in his criteria, he stands a good chance of choosing a suitable candidate to work in the team environment. It is inadvisable, however, for the project leader to recruit someone with the right technical skills in the belief that he can change their personality later; people can modify their behaviour somewhat if motivated to do so, but the limits are often narrow and he should recruit personalities that fit his needs as much as possible. Throughout recruitment, he should have a preference for people who

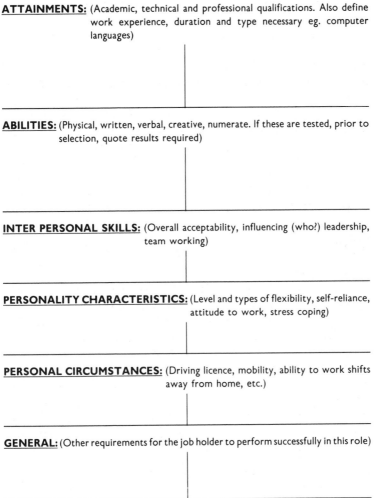

EMPLOYEE SPECIFICATION

ESSENTIAL   **DESIRABLE**

ATTAINMENTS: (Academic, technical and professional qualifications. Also define work experience, duration and type necessary eg. computer languages)

ABILITIES: (Physical, written, verbal, creative, numerate. If these are tested, prior to selection, quote results required)

INTER PERSONAL SKILLS: (Overall acceptability, influencing (who?) leadership, team working)

PERSONALITY CHARACTERISTICS: (Level and types of flexibility, self-reliance, attitude to work, stress coping)

PERSONAL CIRCUMSTANCES: (Driving licence, mobility, ability to work shifts away from home, etc.)

GENERAL: (Other requirements for the job holder to perform successfully in this role)

Fig. 7.2 Employee specification

can accept responsibility and his leadership style, work in a team and, in many cases, influence a reluctant client or user. If the employee specification clearly defines the skills and attributes demanded by and useful for each job to be performed, the project leader will give himself the crucial ability to see whether the candidates' needs will be fulfilled by the job for which they are applying. If individual needs are able to be satisfied by the project, people will be motivated to achieve its objectives.

There are several well-known ways of categorizing information for employee specifications, and the project leader needs to choose the method that best suits him. Whatever the method, he must be constantly mindful that the project team may be together for a matter of weeks or months and that a team may be extended across an organization so that individuals report to more than one boss. If the team is mature and well-motivated, difficult decisions can be made and adverse circumstances coped with by deploying the various talents of its members. Team-building techniques are essential for a project team to work harmoniously and effectively, and the project leader must ensure that people who can respond to these techniques, or who already show evidence of being good team members, are selected. If he is borrowing an employee specification used for another project, he must be certain that it is designed to reject candidates who show no indication of wishing to work within a team or of achieving the sort of objectives the project demands.

If matching jobs with suitable applicants is to be done satisfactorily, the project leader needs a standard way of describing the requirements of the job and the attributes of the people who could carry it out. The two classifications of human attributes in most common use for selection purposes are Professor Alec Rodger's 7-point plan and the 5-point plan popularized by T. Munro Fraser. These two plans are shown in Figures 7.3 and 7.4 respectively. While it is intended that these two schemes could be used to interpret job analysis in human terms, they should be deployed also to set a standard against which individual candidates may be measured. In preparing the employee specification, it is important to consider under each of the headings of the plan:

Essential attributes These are the indispensable attributes the candidate must have if the job is to be performed to satisfactory standards. Minimum technical skills are obviously in this category.

Desirable attributes These are less essential characteristics but may well mean the job is performed to a level higher than the expected standard, or will inspire more confidence or will mean that less training is needed. Experience of similar work falls into this category.

Contra-indications These are attributes which would disqualify an individual. The inability to work away from home when the job requires precisely this would fall in this category.

The strengths of these plans are that they enable the project leader to structure his thoughts on each job and be systematic and objective about

1. Appearance – health, physique; the project leader must remember that the impression given to his clients is important.
2. Attainments – knowledge, experience or qualifications requisite for the job, e.g. 3 years' experience of COBOL programming.
3. General intelligence or other intellectual abilities – testing for these can be considered.
4. Special aptitudes demanded by the job; the ability to pass a programming aptitude test or to be articulate and persuasive.
5. Interests – for instance, certain social interests could indicate a liking for teamwork.
6. Disposition – how important is it that the incumbent be acceptable to others and at what levels? What capacity is required for leadership, accepting responsibility or accepting supervision?
7. Circumstances – will the applicant be prepared to work away from home?

Fig. 7.3 The 7-point plan

1 Attainments – level of education and experience gained are covered and compared with standards set. The project leader should set the basic standards of education needed, but not fall into the trap of setting qualifications unrelated to the job.

2 Innate abilities – the skills, capabilities and personality characteristics the person needs to perform the tasks.

3 Adjustment – since sources of stress are to be found in interaction with others and reaction to work, posts which involve responsibility for other people or considerable pressure demand certain emotional maturity. The selector has the evidence of an employee's past record to indicate the ability to get on with others and subtly exert or graciously accept a certain amount of self-control, and to cope with pressure.

4 Impact on others – the reactions of other people to a particular team member will play a considerable part in the success of that member and of the team as a whole. Some jobs are in the 'front line' and impact is essential. For other jobs, impact is an interesting but less critical characteristic.

5 Motivation – this is difficult to assess and potentially desirable, if not essential, for successful performance. People act and behave to satisfy personal needs, as discussed in the next chapter. When individual needs are satisfied by the set of objectives, interpersonal relationships, environment and management style of the project, there is a good 'fit' and so the individual is likely to feel motivated to work well.

Fig. 7.4 The 5-point plan

selection. Their chief weakness is that they can cause him to be inflexible and perfectionist in his approach when he should be pragmatic. For example, there are times when he should select someone who is only just competent to do a job since a 'high-flyer' could become bored and so perform poorly; he may have to give training to an internal candidate for a job because this is cheaper than external recruitment; he may have to accept a candidate who only possesses the minimum requirements because he is bound by the salary on offer.

In summary, then, the employee specification should allow the project leader to define a candidate's abilities, interpersonal skills and personality characteristics that have enabled them to meet past challenges and reach certain achievements or have caused them to avoid certain challenges or to fail to perform in certain areas. If, for example, candidates have failed to meet the responsibilities of planning their own work or of working alone, but have enjoyed teamwork and performed satisfactorily in a team environment, the project leader may find that they are well-motivated to do a routine, programming job under direct supervision in the team. He may also have a candidate who enjoys supervising and training others and has done this successfully in the past, in which case he will find that they are probably motivated to supervise or train within the team.

7.3 Recruitment methods

Having defined the tasks that need to be performed in order to complete the project successfully and the descriptions of the people needed to fulfil these tasks, the project leader can now decide how to find the people. There are a variety of recruitment sources open to him. He can recruit from within his own company, taking people who are known to him or to his colleagues or, if working for a client, he can recruit from within the client company. He may also contact people who have left his company or who have applied before when jobs were not available. He may also have a staff introduction scheme which rewards employees for introducing potential candidates. Very often, however, the project leader is forced to advertise in the trade press or in the general press, in competition with other employers. This is an increasingly expensive activity in times of general staff shortages and, as an alternative, he may use employment agencies or selection and search consultants. In some cases, it is possible to use government recruitment and training schemes, job centres or even students on degree courses who come to him for periods of work experience. Before deciding which recruitment method to use, he needs to consider the different needs of recruiting a permanent team as opposed to a group of people who will work together for only a short period of time on this one project.

Obviously, it is very often easier and, in many cases, more satisfactory to recruit from within, and the informal methods of building up a team of people from within his company or from the client company, or through contacts and staff introductions, tend to be by far the cheapest method.

If the project leader advertises externally, he needs to take full account of the expense of this method and remember that the ideal advertisement will attract one ideal candidate for each vacancy. Too often, advertisements attract large numbers of applicants and involve him in the expense of sifting out those who are really not suitable. The advertisement needs to be specific and, if he knows nothing about writing good advertising copy, the help of a recruitment or advertising agency will be useful. The content which is generally agreed to form good copy includes a clear description of the ideal candidate, the job title and the description of the job content, the type and size of organization, where the job will be based, what the pay and rewards will be and clear instructions on how to apply. The project leader can also use agencies or recruitment consultants to provide him with temporary, contract or permanent staff. They have access to files of relevant candidates and using these can reduce or eliminate the need for advertising. It may also be necessary for a sensitive project to use the anonymity offered by consultants. In any event, the project leader is likely to use the professional help from a personnel department or personnel adviser to handle this aspect of selecting his team.

7.4 Selection techniques

7.4.1 Short-listing

If short-listing is to be quickly and effectively carried out, potential candidates must get the information they need about the company and the job, and the project leader needs specific information on each applicant, for present and future needs. The advantages of using an application form, rather than receiving merely a letter, are that information will be precise and relevant and can be used easily throughout the selection process and is in a similar form for all applicants.

The project leader should not ask for duplicate information by letter, cv and application form. If he wishes to use the latter, he must send it to the candidate with maximum information about the company and the job, the closing date for the return of the completed form and a warning that testimonials should not be sent yet. He should read through application forms of past applicants and of applicants for similar projects as well as of all applicants for the specific post.

Short-listing from application forms alone is difficult, but there should not be an overwhelming response if the advertisement and additional information are properly drafted. The recruitment method and the right approach are the first selection sifting method. Applications can be compared with the requirements set out in the employee specification and then categorized as 'possible', 'doubtful', 'reject'. If applicants have experience and computer languages relevant to other projects, their forms should go 'on hold' and go towards a resourcing pool, on which the project leader can draw in the future. Out of courtesy, these applicants should be informed that they are being considered for other opportunities.

It may be impractical to interview all the applicants in the 'possible' category and these should be reduced to a practical number—say, five or six for the vacancy and perhaps eight for two similar vacancies—by means of a rigorous check to match essential and desirable criteria on the specification with those of each potential candidate.

Candidates the project leader does not wish to short-list should be rejected quickly and courteously; those who are on the short-list for the next stage in the selection process should be informed equally quickly!

The selection techniques the project leader can consider include an interview with the project manager alone, an interview with a colleague present, group methods, and tests for skills and personal characteristics.

Whether selecting a complete, new project team or choosing an input of knowledge and skills to enhance the abilities of an on-going team, the project leader will be working within constraints as far as finance, time and people are concerned. He must balance the costs and safety of using several methods of recruitment with those of using one method. Rarely would it be advisable to rely solely on the interviews by the project manager alone if he has no experience of interviewing, and rarely would we use tests alone.

Before deciding on who to select, it is worthwhile for the project leader to work out his own management style and motivational needs so that he chooses, where possible, people who are suited to that style. Moreover, an understanding of the effects of his own style of behaviour in the management situation can enable him to be more aware of his impact on candidates when he meets them for the first time or when he interviews them.

7.4.2 Interviewing

It is likely that the project leader will be interviewing all, or nearly all, of the candidates and needs to develop his capabilities as an interviewer. The abilities to question and to listen are not only instrumental to good selection interviewing but are skills which he is likely to use every day in his role as project leader or analyst.

In the selection interview, he is concerned with two things—questioning and listening. The questions need to be open-ended so that he gains as much information as possible; he should not give the candidate the opportunity to answer simply 'yes' or 'no'. The questions should probe the reasons for the answers given; he should say things such as 'what do you mean by . . .' or 'tell me more about . . .' or 'can you give me an example of what you mean by . . .' He can also use link questions to open up new subject areas for discussion. He can, for example, say things such as 'talking about this particular job, how do you enjoy such and such'. He must also be sensitive to all that the candidate says or does; he should register information as diverse as verbal statements, hesitations, nuances in speech, facial expressions, gestures and so on. He also needs to be patient and allow time for the candidate to absorb the point of a question and never assume that he or she necessarily knows what he is talking about. Therefore, he should be precise and clear, and the tone of his voice should encourage the flow of information from the candidate to him.

In terms of listening, the first rule is to stop talking and concentrate on what the other person is saying. The project leader should indicate that he is paying attention and wants to listen, and be patient and not interrupt. He should only speak when sure that all has been said, and the follow-up questions that should be relevant to the conversation that he is having with the candidate. It is important to understand what the candidate means, not just what he says, so he needs to ensure that he understands meanings and not just hears the words.

Having thought about questioning and listening techniques, the project leader is now ready to carry out his selection interview. It is often useful for more than one member of staff to interview each applicant. In some cases, however, this will not be possible and the only assessment of the candidate may be on the basis of one interview. If this is the case and he is unable to have a colleague to help, the project leader should follow the following guidelines.

First, he should prepare the interview as thoroughly as possible so that time available is used to the best advantage. He does this by reading the job and employee specification and the candidate's application form once more just prior to the interview, and should not need to refer to the specification during the interview. He must know exactly what information he is seeking and how he proposes to obtain it; he should decide which areas of the candidate's background need most considered explanation.

Second, when selecting for his team, the project leader must be aware that everyone will have to get along with one another and so might find it useful to interview all candidates within a short time-span, say two or three days, in order that during the interview process he carries an image of all the personalities who might work in the team. Of course, adequate interview assessment notes also help to achieve this.

Third, the clock is a constant reminder that the project leader has limited time in which to confirm and amplify information contained in the application form and to supply the candidate with additional information. In any case, in order to make a fair assessment, he must structure each interview in a similar manner, allow a similar amount of time for them and make his assessment on paper immediately afterwards while his thoughts are fresh and clear. Thus he should leave at least 15 minutes between interviews to make notes which can later be written up formally on an interview assessment form similar to the one in Figure 7.5.

Fourth, the stress of the interview may cause the project leader or interviewee to behave in an unnatural manner. He should establish close rapport with the candidate in order to reduce this level of stress and enable the interviewee to present himself as naturally as possible. This rapport is best achieved if a friendly and encouraging but businesslike manner is adopted.

Fifth, the project leader should structure the interview to get to know the candidate first and then supply details of the job. Throughout the interview, he must relate the candidate's experience and potential to the project in hand. It is no good letting himself be impressed by a candidate because they have strengths in areas not related to the project.

Finally, the project leader should be aware of the halo effect, where a positive or negative impression modifies his behaviour such that he becomes part of a self-fulfilling cycle of increasing positiveness or negativeness. He must also be realistic about what he can offer candidates and not oversell or undersell the job or project.

It is always best for a candidate to be seen by a colleague or team member or a member of the personnel department. This increases the reliability of the assessment, as two assessments are three times better than one. It is interesting to analyse the extent to which assessments coincide and, if they are widely divergent, to think again. When two interviews are decided upon, the interviewers must decide whether to interview together or consecutively. The benefit of the joint interview is that one interviewer has the opportunity to listen and assess the subtleties and nuances of behaviour, voice, gesture etc. while the other pursues their line of questioning. The disadvantage is that it can place greater stress on a candidate and make it more difficult to establish rapport. It might, however, be a useful method of simulating conditions in which a candidate would find himself when trying to communicate with and influence a difficult or important client or user. If interviews are to be conducted jointly, the interviewers must agree aspects to be covered by each of them and the format, and must use the time when they are not speaking to observe the candidate's reaction to their colleague's questions. If separate interviews are to be conducted, interviewers must decide the order, areas of common ground and specific items to be covered by each of them. If there is a professional personnel department, they can assist with the interviewing or give guidance about it; they can also help to assess an in-house candidate, as can colleagues. Knowing someone well is a great advantage, but it must not be allowed to cloud objectivity either for or against them; the requirements of the project are paramount.

In any event, whether interviewing separately or with a colleague, there are a number of things a project leader needs to do if the interview is to be successful. Before the interview, he needs to make the room comfortable and remove any barriers between the candidate and himself. He should, for example, check that there will be no interruptions during the interview and make sure that there are some writing materials at hand for note-taking and that the candidate's application form is in front of him. If it is intended to give the candidate brochures or information about the company or project, the project leader should have these to hand too.

At the beginning of the interview, the project leader should introduce himself and explain that he would like to know more about

JOB INTERVIEW ASSESSMENT

Candidate's Name .... **Date Interviewed** ....

Job Title ....

Factor	Comment	Evidence, e.g.	Rating

Attainments
 Educational/professional
   qualifications
 Other training
 Jobs held
 Responsibilities
 Achievements – technical/
   business
 Technical experience
 Interests			

Innate Abilities General intelligence Special aptitudes – written/ verbal/numeracy/creativity Learning ability Planning/controlling			
Circumstances Location/travel Family situation Availability/notice Current pay			
Interpersonal Skills – *Impact on Others* Physical Verbal Disposition Acceptability Influencing skills Team orientation Leadership			
Motivation Drive Enthusiasm Self-reliance Attitude to work Emotional adjustment Stability			
Overall Assessment			

Fig. 7.5

N.B. Rating: 5 = Excellent (Top 10%); 4 = Above Average (20%); 3 = Average (Mid 40%); 2 = Below Average (20%); 1 = Very Poor (Bottom 10%)

Offer/Hold/Reject   Interviewer's Name ....

Signed ....   Dated ....

Fig. 7.5 – (contd.)

the candidate, who may ask questions later. He should also explain that he is going to take notes, if he intends to do so. The 'ice-breaker' statement or question with which he is going to begin the interview should be prepared carefully beforehand. The first few minutes of an interview are essential in establishing good rapport with a candidate and stress should be avoided, since truth is much more likely to flourish in a relaxed atmosphere.

The interviewer should spend more time listening than talking and should encourage the interviewee to take as active a role as possible. Long pauses left between questions, therefore, allow the candidate to explain himself or herself to the full. The project leader should control the interview by asking the questions, summarizing and making encouraging remarks. Questions should give the impression of moving naturally from one subject to the next. Throughout the questioning and listening, he should ask himself whether the evidence indicates that the candidate has the requisite knowledge, experience, skills and personal characteristics essential for the job. Does the evidence he is getting indicate other attributes which are desirable for the job? The project leader should ask the candidate to elaborate upon his strengths and weaknesses and see if their opinions coincide. It is also useful to see how candidates cope with questions, as well as what they actually say; the project leader may, for example, need to select someone who is able to deal with a particularly difficult client or to lead a particularly fragmented team. Whatever the circumstances of the project, he can therefore gear his questions not only to get the facts but also to see if the interviewee can persuade him or sell himself during the interview. He should always use qualitative, open questions and assume that omissions need to be explored and should, therefore, guide the interviewee back to the points of interest if straying from them. He needs to ask the candidate what was better, different, started or stopped in his present job because of his own activities and avoid making early judgements and build assessment of his capabilities on the track record demonstrated. The project leader should also ask the applicant what sort of references he will get.

The project leader wants his team members to be motivated to achieve the team's goals and therefore needs to find out what goals people appear to be pursuing now, why they are pursuing them, and what goals are not of interest to them. What people like most does not necessarily coincide with what they do best. However, when people are doing not only what they enjoy but what they do well, there is a powerful combination of motivation and abilities. When drawing out someone's motivational needs, a suitable framework for questioning would cover their past work, their present work, their hopes for the future and how they spend their leisure activities. When asking them about their previous work, the project leader should aim to uncover the jobs they have most enjoyed and the work they have enjoyed least. He should get them to give an example of a working situation which made them feel good and one which made them feel bad and get them to explain the reasons for these choices. In their present job, he should find out what they really like about it and what interests them particularly in it; and also attempt to explore why they want to change. As far as the future is concerned, he would like to know what their plans are for their working future, what aspirations they have and what particular aspects of a future career might not appeal to them. Getting to know about their leisure time can tell him whether they are a team person or a loner, and he can begin to build up a picture of how they might fit into the team.

Where an applicant has already demonstrated clear achievements in his working life, there is good evidence of him being motivated to achieve particular things. However, while motivation towards other things provides some clue, it is not sufficient to enable the project leader to tell with certainty if a candidate will be motivated to his particular job or project. If he is selecting candidates from his own company, he can find out about their motivaiton by day-to-day observation and by asking questions, but needs to be aware that the needs of each project differ and that people change over time. When he has no evidence to go on, perhaps because he is selecting junior staff with little working experience, the interview needs to be structured to draw out the past and to identify the challenges they have confronted or avoided, and the tasks they have seen through, perhaps in their education or training. The framework of a typical interview, showing the flow of information between interviewer and applicant, is shown in Figure 7.6. In summary, then, throughout the interview, the project leader should make good eye contact with the interviewee, listen, smile occasionally, and leave him with a positive impression of the job, the project and the company. Even if he doesn't want a particular applicant, the project leader should

A = Interviewer   B = Interviewee

A→B Introduction – settling the candidate, explaining who you are, how long the interview will last and what else will happen during the visit. Very brief comment on vacancy and company.

B→A Analysis of candidate's history – beginning at the earliest relevant point progressing forward – education, early jobs, recent jobs. Reasons why? Career path?

B→A Analysis of candidate's present situation – present job and personal circumstances, reasons for being on job market.

B→A Analysis of candidate's future – what they want out of their next job, career plans, general aims in life.

A→B Discussion of vacancy – the interviewer matching the picture of the candidate against the specific requirements of the job, the candidate finding out exactly what is involved.

A→B Conclusion – terms and conditions etc. The interviewee has remaining questions answered and is advised of what will happen next.

Fig. 7.6

try to complete the interview with him so that he is impressed and wishes that he could have joined.

If, during the interview process, the project leader believes that the interviewee is unsuitable in some way, he should first question himself about this belief or feeling. It could be the 'halo effect'—his reaction to something said or a mannerism that has personal significance for him. He should test out his feelings and put his beliefs to the test: this can be done by asking more questions, by probing until he has put the initial feeling to one side or confirmed it with factual evidence.

The project leader should never abandon interviews in a demotivational way. He must keep an open mind—some people take a while to warm up and it takes a while to get to know them.

7.4.3 Group selection methods

There are two other selection techniques worth discussing here. The first deals with group procedures where the candidate meets other members of the team. These procedures are most useful for team leader positions or where jobs require high interpersonal and persuasive skills. When selecting candidates for an established team, it is also helpful to spend some time talking informally with team members to see how they communicate with them and whether the team members feel that they will all get on together. When building a team from scratch and selecting the leader, the style of leadership should first be ascertained. Usually, a consultative rather than a coercive style will be looked for, and one which is supportive rather than authoritarian. Observers watching groups participating in group exercises or problem-solving should be well versed in the job that the leader will have to carry out and in the employee specification, and should be trained to recognize the behaviour symptomatic of the required leader. Techniques of this kind are often used in military applications, but they are unlikely ever to replace interviews and are generally used to supplement the information gathered at them.

7.4.4 Selection tests

Selection tests can be a useful supplementary source of information about interviewees, provided they are chosen and interpreted by trained people. They check against the subjectivity of the interview and ensure that the candidate possesses the skills or personality to do the job. They also ensure that maximum information is available before the final selection is made. When a high level of skills is not needed, these tests can be used to compensate for candidates who find interviews too stressful. There are a number of different types of tests: intelligence tests measure thinking power, clarity, speed of thought, and a commonly used one is the Alice Heim AH5 test. Aptitude tests indicate particular abilities, such as programming aptitude, and the Computer Programming Aptitude Battery is well-known in the selection of programmers. Achievement tests show knowledge and proficiency in particular areas and personal inventories predict the way in which the candidate will tend to behave. There are personality factor tests and occupational personality questionnaires, of which the best known one is probably the 16 Personality Factor Test—often referred to as the 16 PF.

If using tests, the project leader needs to be sure that those used are consistent in their measurement so that a candidate who does the test twice should achieve similar results. The test must be valid in that it should measure what it claims to measure, and computer tests which are indicative of past training rather than predictive of potential and therefore future performance are not valid.

7.4.5 Final assessment

It has already been said that the project leader must make and record a preliminary assessment after each interview and should also do this after each group procedure. Assessment is best done in the job interview assessment form shown in Figure 7.5, so that all candidates are rated in the same way and a clear comparison can be made between candidates and the employee specification. When completing the form, the project leader should assess only those attributes listed as essential or desirable on the specification; if no requirement has been listed, no assessment is necessary and that requirement must be ignored. He should rate the interviewee only where there is *evidence* to support the assessment. He must avoid assuming that a candidate can perform well in one area because there is evidence to show they have performed well in another area.

The project leader compares the assessments for each candidate with the requirements of the employee specification, focusing first on that which is essential and, secondly, on that which is desirable. Information received after the interviews, such as results of a medical or a selection test, should be considered as well. Test results may either set a minimum acceptable standard below which candidates are automatically rejected or provide additional or confirmatory information to the interview assessment.

The project leader should try and have a second candidate in case the preferred one turns down the offer. When the offer is made, the project leader should ask the candidate if he may approach past or present employers for references if the offer is accepted. References from present employers may be fresh but they can also be biased; telephone references followed up by written confirmation are the most reliable. When selecting in-house candidates, he has the more reliable opinions of the colleagues whom he trusts on the candidate's files as the result of appraisals or staff assessments.

Offers should be made immediately a decision is reached. It is often useful to telephone the successful candidate and follow up the conversation with a letter of appointment and a formal contract to reach them as soon as possible.

7.5 Summary

It is necessary to evaluate selection techniques for future use: this can be done by assessing the performance of individual members of the team as a whole at stages in the project and at the finish. The project leader can see whether the team is working effectively and where its weaknesses lie; he can encourage team members to mix informally to get to know one another and discuss points of mutual interest or problems with the project. He can meet members formally from time to time to show them the direction in which the team is going. Just as he needs feedback on his performance, so his team members need feedback on theirs. It is no use selecting a perfect mix of people and then resting on his laurels: just as he has chosen them, they have chosen him and he has a commitment to them.

8 Motivating and leading the team

8.1 Introduction

People strive to satisfy their needs; everyone needs some job satisfaction and we all have different needs which vary over time. Needs produce the personal motivation to satisfy the needs themselves. We may understand intellectually that people at work need to feel motivated to achieve goals, but only with practical experience can project leaders create an environment in which self-motivation of an individual contributes towards the achievement of the project's objectives. If we are to understand the motivational needs of others, we need to understand our own motivational needs first. There is a simple model we can use to help us think about ourselves and about others. On a blank sheet of paper, we draw two lines across the page, dividing it into three from top to bottom. In the top section, we write a list as long as we can of our needs at work; in the middle section, we put down our feelings when each of these needs is unfulfilled; and in the bottom section we explain how we behave when we feel each need is not satisfied. If, for example, one of our needs at work is companionship, we feel lonely, miserable, no longer interested in the job if our need is unfulfilled and in consequence our behaviour becomes withdrawn or uncaring. This is shown in Figure 8.1.

Understanding his own needs, and his likely feelings and behaviour when they are not met, helps him to understand and plan for the situations that will arise during the project when he may not be able to satisfy his own motivational needs. In particular, he should be clear about his needs in relation to managing others, for unless he has them clear in his own mind and is able to cater for them, compensate for them or cope with their absence, he will not function properly as project leader. As we saw in the previous chapter, it is important to think about need fulfilment when recruiting people, since the greater the likelihood

Examples.

Section 1 (needs):	companionship	independence
Section 2 (feelings if unfilled):	lonely, miserable no longer interested in the job	trapped, angry, sad, used
Section 3 (behaviour when feeling dissatisified):	withdrawn, uncaring	aggressive, leave, go sick (real or not)

Fig. 8.1

of their needs being met, the greater will be the chances of the project being successful. When he knows what the motivational needs of the team members are, the project leader can structure the project and allocate the tasks so as to satisfy as many individual needs as he can, consistent with meeting project goals. This means that each team member is motivated to do their best with the tasks that are assigned to them. Additionally, he can be clear in his mind when the needs of the project require him to allocate tasks with which people feel uncomfortable and can then take special steps to allow for this.

8.2 Motivation

There are two commonly-used motivational models which are worth examination: those developed by Maslow and Herzberg; the relationship between these two models is shown in Figure 8.2. According to Maslow, people only feel motivated towards satisfying their needs at any of the levels in the hierarchy once the lower-level needs are satisfied. Herzberg believes that the job content will motivate people to work, but only when the context of the job is satisfactory. His hygiene factors – the job context – are not satisfiers in themselves but make people dissatisfied when they are not properly present. Both theories help us to understand how to get the most out of people, and their generalized statements provide a conceptual framework within which to manage in practice.

When someone says they feel motivated, they say they are 'turned on'; and when they are de-motivated, they say they are 'turned off'. The project leader's job, therefore, is to provide as many 'turn ons' as possible. Fortunately, there are many opportunities to do this since most of them come from decisions about structuring the job, appraising the members

Highest

The hierarchy of needs (Maslow)	The motivation/hygiene theory (Herzberg)
5 Self-fulfilment needs The self-image realized potential	Motivation factors Achievement Recognition Work itself Responsibility Advancement
4 Ego needs The desire for esteem	
3 Social needs Interpersonal activities Group experience	Hygiene factors Company policy and administration Supervision – technical Supervision – interpersonal relationships Salary Working conditions
2 Safety needs Protection against risks – security	
1 Physiological needs Air, food, shelter, warmth etc.; today these are satisfied by money	

Most Basic

Fig. 8.2

of the team, and rewarding them or promoting them if they are successful.

The de-motivating factors tend to fall into two areas, pay levels and physical environment, and the project leader is typically unable to do very much about these. Significantly, however, his management style and the working climate he produces can also be de-motivators if not appropriate to the team he is managing or to the work in hand.

In summary, people work best when the following things have been taken into account:

(a) the nature of the job – it needs to be interesting and stretching and give responsibility for a whole unit of work rather than for many fragmented and unrelated tasks;

(b) achievement – if the team members' goals can be identified with those of the project, then more is achieved. Success breeds success, and recognition of it by management promotes further success;

(c) advancement – here we feel that once we have achieved our goals, we shall be able to move on and develop, learn, or gain status and so on;

(d) company environment – we work best when employed in a happy, open atmosphere, and where the physical environment is not unpleasant.

There are also a number of corresponding de-motivators: if the nature of the job is impossible, or too easy, or too routine; if the level of support or guidance is inadequate; if there is inflexible control or negative criticism, or lack of recognition: these things are all symptomatic of unsympathetic or bad management. The lack of responsibility and restrictions on the freedom to act, together with an unfulfilling job, mean that team members are turned off and, if they are working in a closed or threatening environment where people lack confidence to accept change or to express constructive criticism, then there is little or no communication, and team members feel de-motivated towards achieving the project's goals.

8.3 Achieving motivation

When considering what to do in practical terms in each area of motivation, the project leader needs to bear in mind the different and contrasting needs of team members which may themselves alter over time. If, for example, someone achieves status or moves to a different task, then their needs change and the project leader needs to recognize this. If the project leader has done a good job in selecting his team, he should not have to deal with too many conflicting needs; but he may not always be in a position to choose all the members and will, therefore, never be working with a 'perfect' team, only with a potentially successful or unsuccessful one. Nevertheless, in order to create a happy and motivational environment, it is up to him as project leader to provide as many of the 'turn ons' as possible, and to avoid whatever de-motivates his particular group of people.

Since the main motivator for staff working on a systems project is almost always the nature of the job itself, the project leader needs to pay particular attention to this. He needs to get to understand everyone's work-based needs and to manage the work content and context accordingly. He should be observing, listening to and giving feedback to his team so that they understand how they and everyone else is performing, who is under pressure and who may need help. At the start of the project, he needs to communicate the project aims and objectives to the whole team, and specific objectives to each individual; he can do this by having a formal meeting with the whole team, and individual meetings with the team members. If he has selected well, he will have a potentially well-balanced team of people, some of whom will wish to take on as much responsibility as their tasks permit and others who will prefer to be supervised more closely. This will give man-management opportunities for those who are ready for it. Normally, team members will be technically competent to do their tasks, and these should be made optimally interesting to them by the use of a number of man-management methods. The project manager can, for example, so structure the job that individual members of the team are given specific responsibility for delegated tasks. These could, for example, be as follows:

(a) people – responsibility for or involvement with the assessment, training and development of junior staff;

(b) finance – responsibility for individual budgets, expenditure authority and the requisitioning of materials and services;

(c) planning – allows involvement in strategic decisions and in the selection of work, the organization of schedules and priorities, and in the freedom to follow up their own ideas wherever possible;

(d) actions – makes voluntary and verification or checking of work, allows an authorship of letters and reports and encourages informal links between all sections of the project, and between the team and client;

(e) communications – provides all necessary information and cost data as often as needs require; clarifies individual's and group objectives and gives recognition of achievement and precise information about where performance needs to be improved;

(f) training – encourages new techniques and their cross-fertilization within the team; trains in interpersonal skills, management techniques, budgetary control and technical matters wherever possible.

The project leader can also make the jobs of the team members as satisfying as he can by making them as complete as possible and letting individuals in the team see through specific tasks from start to finish. It is also useful in the interest of efficiency to allow the members of the team to make changes in the way work is done in the interest of efficiency, and to allow for personal innovation and creativity wherever it is possible. After all, people are most committed to their own ideas and methods.

It is also useful to ensure that all jobs have variety and that job-holders are given authority to take decisions that fit within the remit of their work. Sometimes it is helpful if people are allowed to share or to swap jobs in order to enable them to gain wider experience and to get a better appreciation of the scope of the whole project. If the project leader has too much work to handle, he may be delegating insufficiently so that he is becoming de-motivated by overwork and de-motivating his staff by lack of variety and responsibility. When he does delegate, he has time to focus on the management aspects of his role as project leader and to develop subordinates. He needs to ask himself 'What do I do that could be done by someone else on the team?' Repetitive jobs can be delegated, such as those requiring frequent, minor decisions and those entailing preparation of time-consuming details.

Team members will feel most motivated when given further parts of the job they do already, when given special jobs to handle in their own way and when their usefulness is being increased by their job experience. Usually there are team members keen to learn and try out new skills and others who are keen to improve their existing knowledge and skills. When delegating, the job needs to be defined in terms of its objectives and time-span, but the method can be left to the person concerned. The limits of authority being delegated need to be clearly defined as well. A job may be done without reference to the project leader, for example; or it can be done with the project leader being informed once action has been taken; or action may be taken following consultation with the project leader. Different circumstances will, of course, allow different levels of delegation. The main thing is to be clear about the amount of authority being delegated because, in the final analysis, the project leader stands accountable for the tasks that have been delegated and must be supportive when things do not go according to plan. It is important for the project leader to remember that when delegating, he should consider how to gain optimum benefit for the usual job-holder and the delegatee; one person's job enrichment could be another's job impoverishment! Throughout the project then, the project leader must keep in perspective the individual needs of his team members simultaneously with the needs of the project; no one finds this easy. Just as a mature individual will be highly motivated to take on new tasks with little direction or support, so a mature team will be appropriate for a 'delegating' style of management.

Once he knows all his team members well, the project leader will be aware of the varying amounts of recognition and reward which they need; some people may find public recognition and reward an embarrassment; others may seek only sufficient reward for doing their job well; and for others, recognition and reward are high on their need priorities. However, people usually do thrive on positive feedback when they are performing well and often welcome positive criticism when they are not. Weekly informal reviews with individuals or small groups, informal group meetings, briefing meetings and general progress meetings should be part of a team's way of life. By appraising the work of individuals or groups, the project leader should be able to help people through their difficulties and have an opportunity to reward effort and contribution above the norm. In weekly briefing meetings, for example, he can talk about the direction in which they are moving and discuss any hold-ups, thereby encouraging people to help one another and to identify with the team. In regular progress meetings, this can be done more formally and so these should be planned carefully to include all elements of the project. At all meetings, he should encourage maximum participation in order to promote team spirit and also to gain knowledge of the team and its work. In the previous chapter, the importance of questioning and listening skills were discussed, and it is crucial here that the project leader use these skills in his daily communication with individuals and groups at meetings.

Real communication is concerned with the development of genuine understanding and the sharing of problems. Therefore, feedback is vital in this process and is most useful when:

- it takes into account the needs of the receiver as well as the giver;
- it concentrates on stating events or feelings rather than evaluating them;
- it concentrates on things which the receiver can do something about;
- it is specific;
- it is near to the event at a time when people are receptive;
- understanding of it is checked.

Many people's reward will be attaining the goals set for the project. The project leader should therefore underline the extent to which he is getting there and take every opportunity to convey his enthusiasm for the joint venture. In times of stress, such as when coping with an uncooperative user or with illness or system changes, he must maintain the keen interest in the team and listen carefully to its difficulties and reward it when it copes successfully. While some team members may resent individual group meetings as an intrusion of their work – and the project leader must be aware of this – the majority will be encouraged by the concern and involvement. Where reasonable, possible and advantageous, monetary rewards can be used to consolidate the recognition process, but it is unwise to try and use money too often as its absence then becomes a de-motivator on more and more occasions.

Achievement

At the onset of the project, the project leader should aim to give everyone the opportunity of early and regular achievement by structuring the work appropriately. He should then show individuals how their own goals can be identified with the team's goals, and at the same time give them a chance to realize that when the team performs well, this is their personal achievement too. At informal and formal meetings, he can praise achievement. Where people are under-achieving, he can appraise their tasks to see what can be done to make these more stimulating, perhaps by re-organizing tasks or providing extra training where necessary. If his style is to be open with his people and to offer positive criticism, then he knows that they will come to him when they are dissatisfied with their own or the team's performance.

Advancement

Opportunities to promote staff within a project team come infrequently. It is important, however, for team members to know that their performance is being fairly assessed so that, when promotion opportunities do arise, high achievers can be rewarded. More often there is the opportunity for the individual to develop his own skills and knowledge through the work of the project. This continuous learning climate is found to be highly motivational and is a material form of advancement and, as a positive motivator, is entirely within the project leader's ability to organize.

Company environment

While some factors will be outside the project leader's control, team members need to know that company reward and recognition of the team's contribution will be given. Bad company rules, pay structures, physical working environment and overall poor company communication systems can all be de-motivators. The project manager will need to insulate his team from them if this is the case.

Working through all of these elements and central to the successful management of the project will be the project leader's qualities of leadership. The three important elements in this leadership will be his ability:

- to lead by example
- to organize work and develop his staff
- to demonstrate and generate enthusiasm for the project

The day-to-day behaviour and actions of the project leader have a material effect on the performance of the team, and different leadership styles affect teams differently. It has been suggested that leadership varies along a continuum, from being 'boss centred' at one extreme to 'team centred' at the other. Four clearly distinguishable styles along this continuum are those of a project manager who tells, sells, consults or joins.

The project manager who tells is one who makes his own decisions, announces them and expects them to be put into effect without question; the manager who sells, still makes the decisions himself but convinces others of their rightness. He differs from the extreme autocrat only in so far as he persuades his subordinates to accept his decision rather than expecting them to accept the decisions automatically.

A consulting manager is one who does not decide until he has presented the problem to the group, listened to their suggestions and formed a conclusion which is then the decision that the group has to follow. The final style of leadership is where the manager joins the group which makes a joint decision. The manager defines the problem and the decision taken will reflect the majority opinion in the group.

Managers who tell are thought of as efficient and decisive, though autocratic. Managers who join appear to be most fair and supportive but perhaps lacking the ability to give decisive leadership. The consultative style is the one typically thought to be strong in efficiency and in human criteria. Managers who consult seem helpful and keen, able to build team spirit, to delegate, to solve problems and to develop forward plans. At the end of the day, there is no one style which is right or wrong in all circumstances, and project managers typically learn to pick the most useful parts out of each and to apply them in the appropriate circumstances. There are situations such as crises when the time is short, or where there is no clear answer to a problem when decisiveness and the telling approach is most appropriate.

It is important to think through the different kinds of situation that can arise during a project, and to use the style which will be most appropriate in each situation. However, much as project leaders can modify their own behaviour and actions, they do need to remember that they too have personal needs which will affect the range of leadership styles that they are able to adopt. Listed below are seven descriptions of different leadership styles, and it is worth a project leader considering which of these most aptly characterizes their own personal needs and style. Certain characteristics to be found in most good leaders and managers in the workplace are found on the positive side of all of the following different styles:

(1) *From critical dictator to informed critic.* Critics can discuss, judge, discern and communicate but have a propensity to become dictators who rage, repress and insist on things going their way, who resist change and new ideas. The positive side of this style is with the informed critic who, without making judgements, waits until he has all the facts. Informed critics are good listeners and work well under pressure, expect much from staff and so tend to get it. They set limits so that people know what is expected and so provide a sense of security.

(2) *Benevolent dictator to supportive coach.* The benevolent dictator will overwhelm and manipulate staff with concern, insist on things going their way and thus encourage dependent responses. Underneath the paternalistic mask can be the condescending attitude of someone feeling threatened by another's talents. They will make such remarks as, 'let me finish that for you – you've been overworking!'. A more positive use of this caring personality trait is found in the supportive coach who encourages staff to develop to their full potential and is pleased with their successes. They are competent at listening, giving feedback and acting with support and sympathy. Therefore, people tend to feel understood by the coach and often respond with higher motivation and prosperity, although some may resent what they see to be an over-protective boss.

(3) *From loner to liberator.* The loner is neither critical nor nurturing. Loners stay uninvolved, conveying a 'don't interfere with me' message, and are unconcerned over the well-being of their staff. This behaviour may well be exhibited by a technical or subject specialist when put in charge of a team for the first time. On the positive side, the liberator is not overly helpful or repressive and, as they expect staff to be creative and competent like themselves, they often get this response. A self-motivated employee will thrive in this atmosphere.

(4) *From computer to communicator.* The over-rational analyst continually processes data – on products, numbers and people – and can be inadequate in dealing with the human aspects of management. Computers rarely show concern or sympathy and will say 'let's not discuss this personal problem, let's get on with the job'. The positive side of this mechanistic computer-like trait is a responsive communicator, whose ability to collect and analyse data and estimate problems is humanistically oriented. Their clear and effective communication skills are appreciated since their staff know precisely where they stand.

(5) *From milksop to negotiator.* The compliant, agreeing milksop leader is a doormat who tries to please everyone and ends up pleasing no one. Such leaders are fearful of conflict and will avoid giving their team the chance to mature by battling through problems to reach a solution. This pacifying nature can have a positive side, however: when pacifiers learn to assert themselves, they can take on the role of a fair negotiator, who can provide a balanced working environment. Because of a seeming willingness to examine the other side of most disputes, people feel they can relax and let down their defences in the presence of a fair negotiator, so that there is more opportunity for conflicts to be talked through and problems understood, even if not solved.

(5) *From punk to partner.* The aggressive, rebellious punk is out to win by any means, fair or foul. This kind of leader has so much hostility that it will be exhibited in such behaviour as procrastinating when discussing personal problems. When hostility is open, it may be expressed as an exclusive competitiveness, which permeates the whole environment. When things go badly, the punk seeks revenge with exclamations such as 'tell the client I don't care about his business!'. Partner leaders have a strong, positive value when they channel their assertiveness towards being a partner in teamwork. They will fight for success, share information with their staff, direct them in new techniques and so develop a positive team spirit.

(7) *From scatterbrain to innovator.* Scatterbrains can be too immersed in creativity and experimentation even to communicate their new ideas. They may lack the necessary communication skills to enable new ideas to be adapted and lack the requisite managerial skills to cope with the daily operation of the team. Their flair for creativity is positively expressed in the innovator, who provides an exciting working environment, once they have got to grips with the day-to-day management of the team.

Such leaders are well liked for their ready flow of fresh ideas and enthusiasm that energizes others. They tend to be receptive to other people's ideas, so that the team's creative potential is fully realized. Whereas some people cannot cope easily with the pace of an innovator, always seeking fresh ways to do things, others find this stimulating.

8.4 Developing the team and team effectiveness

The potential of the team to reach maturity will depend largely upon the qualities of its project leader. If he provides the right motivational framework, he will promote the team's effectiveness as an achiever of its objectives and as a problem-solving unit. The following checklist can be used throughout a project to enable him to bear in mind all the major ingredients that go towards making a happy and efficient environment and to realize when they are lacking:

Task effectiveness
   (1) Does the group use a methodical approach?
   (2) Are human and material resources used effectively?
   (3) Are activities and ideas co-ordinated by him, by someone else or through a plan?
   (4) Do members seek and offer information and ideas?
   (5) Are ideas expanded and tested in the team?
   (6) Is action initiated-and undertaken with energy?

Group climate
   (1) Are members encouraged and supported by him and by one another?
   (2) Are their contributions valued?
   (3) Are members brought into discussions?
   (4) Does the team set standards for itself to use in choosing procedures and assessing decisions?
   (5) Are personal issues given voice and dealt with?
   (6) Do members thoughtfully, rather than begrudgingly, accept group decisions?

Destructive behaviour
   (1) Are any members highly aggressive, competitive, withdrawn in behaviour and actions?
   (2) Do members disrupt the work of the team in any way?
   (3) Do members exhibit self-oriented behaviour at meetings by bringing up issues for personal reasons rather than because they relate to the project?

The use of this kind of checklist and the project leader's own day-to-day observations of what is taking place in the life of the team are very important. Inevitably, there will be opportunities where improvements can be made and Figure 8.3 is a useful model.

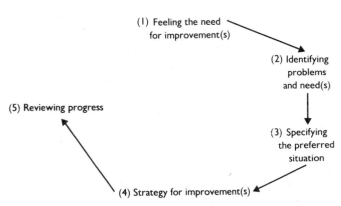

(1) Feeling the need for improvement(s)

(2) Identifying problems and need(s)

(5) Reviewing progress

(3) Specifying the preferred situation

(4) Strategy for improvement(s)

Fig. 8.3

Review meetings can also be used to assess whether or not project activities are running smoothly. The following simple six-step plan helps in this process (Figure 8.4):

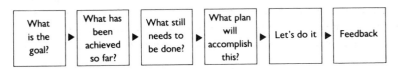

| What is the goal? | What has been achieved so far? | What still needs to be done? | What plan will accomplish this? | Let's do it | Feedback |

Fig. 8.4 Six steps towards good review meetings

At step one, the project leader considers once again the objectives and the targets to be achieved in order to accomplish the main project goals.

At step two, he discusses all tasks that have been completed successfully so as to give everyone a clear idea of the required end performance and enable optimum energy to be devoted to accomplishing the outstanding tasks adequately with the best possible use of time and resources.

Step three is concerned with enabling the team to piece together all the relevant information and to see what is missing. An open discussion about feelings, facts and achievements is important at this step.

In step four, with the help of the team, the project leader can clarify a broad plan and the specific methods to be taken by each individual in order to achieve it.

Step five is concerned with action, and tasks should be completed as specified in the plan. A well-organized team can galvanize itself into action to complete each task within time-limits and under pressure.

At step six, team members learn from seeing the results from their actions, as the project leader can do himself, and feedback information should be collected, clarified and shaped into new operational rules. This feedback information is very important as without it there is little chance of further improvements being developed.

It is inevitable that conflict will arise during this process and, while in itself it is not undesirable since it is only through the expression of differences in the team that good problem-solving can take place, the project leader must be sufficiently vigilant to realize when this conflict is having a negative effect and when differences between individuals or between his group and another group are proving to be counter-productive. Negative effects can be reduced by finding an over-riding goal which is accepted by all team members as being imperative, so that a win–lose situation between two members of the team does not occur since both are united to achieve that goal. Encouraging feedback between people in conflict with each other also tends to reduce the negative effects of this, because by encouraging feedback the individuals clarify the situation and build a more common language. As the team matures, members' needs become integrated with group needs and there should be less self-orientated behaviour and greater focus on the tasks to be done.

An effective team will therefore manage complex situations creatively and, because its resources are being developed, it will respond energetically and rapidly under pressure and will take care to clarify the individual competence of each team member and gain excitement from this personal growth. This excitement and energy will feed the team as a whole. People will also have confidence in one another's skills and also know when to give and receive assistance. Mature teams are capable of making better-quality decisions than most individuals because of the scope of skills and experience they possess. They are also able to make joint decisions and consequently their commitment to the project is higher. Individuals often feel it is difficult to influence a large organization, but in a team they are able to see the impact they have and can feel a sense of belonging which 'turns them on' to the team and its project. The potential to reach this sense of maturity depends on the extent to which the project leader can create an environment which is motivational.

Finally, there is a model of group development that characterizes the stages from getting together to becoming an effective happy team. This model is called Cog's Ladder and is described below and illustrated in Figure 8.5.

The Cog's Ladder model consists of five steps in the development of groups. In the first phase, called the polite step, group members get to know each other, share values and establish the basis for a group structure; on this step, group members need to be liked. The second step is called 'why we are here': on this step, the group members define the objectives and the goals of the group. The third step consists of a bid for power: on this step of the ladder to maturity, group members attempt to influence one another's ideas, values and opinions, and this stage is characterized by competition for attention, for recognition and for influence. The fourth step is co-operative – it is called the constructive step. Here, group members are open-minded, listen actively and accept

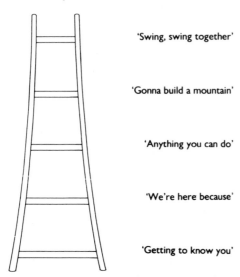

'Swing, swing together'

'Gonna build a mountain'

'Anything you can do'

'We're here because'

'Getting to know you'

Fig. 8.5   Cog's Ladder: a model for the development of groups

the fact that others have a right to different value systems from their own; this step is often also referred to as the 'team action' step. The fifth and final step is one of unity, high spirits, mutual acceptance and high cohesiveness; it is called the esprit de corps stage.

In the polite stage, the initial item on every group's agenda is to get acquainted, whether or not the leader of the group allows time for it. Generally, this begins with members introducing themselves to each other, and name tags may be provided to facilitate this 'getting to know you' process. Polite conversation includes information sharing which helps group members to anticipate each other's future responses and behaviour within the group. During this phase, some group members often stereotype one another to help them categorize the fellow members of the group. The group establishes an emotional basis for future group structures and cliques are formed which will become important in later steps. The items on the hidden agendas of group members stay hidden on this step and do not usually affect behaviour at this time; this is because the need for group approval is strong. Conversely, the need for group identity is low and may be completely absent. Group members tend to participate actively, though unevenly, and usually agree that getting acquainted is important to the group; consequently, conflict is usually absent at this step. The rules of behaviour during this early stage of group development include the need to keep ideas simple; to say acceptable things and avoid controversy; to avoid serious topics and disclosure of hidden agendas.

When a group is ready to grow beyond this polite step, it usually enters the 'why we are here' stage. Group members want to know the group's goals and its objectives and some demand a written agenda. It is at this second step that cliques begin to wield influence: they grow and merge as clique members find a common purpose and hidden agenda items begin to be sensed as group members try to verbalize group objectives most satisfying to themselves. Identity as a group is still low; the need for group approval has declined from what it was at the previous step as members begin to take risks and display commitment. There is usually active participation from all members of the group at this time.

The third step on the ladder, the 'bid for power', is characterized by competition. Group members here try to rationalize their own position and to convince others in the group to take the action that they feel is appropriate. Other members are, however, closed-minded and are accused of not listening, which means that conflict rises to a higher level than at any other stage of group development. A struggle for leadership may occur involving active participation by all cliques or sub-groups, and typical attempts to resolve this struggle include voting, compromise and even seeking arbitration from outside. The group does not feel a strong team spirit here and some members may feel very uncomfortable as latent hostility is expressed. Indeed, some members who contributed willingly at earlier stages remain completely silent in the 'bid for power' phase. Other members, however, relish the opportunity to compete and attempt to dominate the group. The group does not build an identity in this phase and, unfortunately, some groups never mature past this stage. They can, however, fulful their task even though research evidence shows that solutions coming out of groups at this stage of activity are not optimum solutions in that they never satisfy all group members and, at best, are products of compromise.

To reach the fourth stage – constructive stage – an attitude change is required where group members give up their attempts to control and substitute an attitude of active listening. It is in this stage that group members are willing to change their preconceived ideas or opinions on the basis of facts and evidence presented by other group members. Individuals will actively ask questions of each other, team spirit starts to build with a consequent dissolution of cliques, and real progress towards the group's goals becomes evident. Group identity now begins to be important to the group members, and when conflict arises it is dealt with as a mutual problem rather than as a win or lose battle. Because of the group or team's willingness to listen and to change, activities in this stage often use the talents of any individual who can contribute effectively, and practical creativity can be high because the group is willing to accept creative suggestions. Furthermore, creative suggestions are actively solicited by the group, listened to, questioned, responded to and, if appropriate, acted upon. It is likely that an optimum solution or decision reached at this time is almost always better than that offered by a single group member. It is for this reason that some businesses are attempting to organize for team or group activity rather than for individual activity.

The final and fifth step on the ladder of group growth is the esprit phase. It is here that the group feels high group morale and an intense group loyalty. Individuality and creativity are high and members of the group participate as evenly as they ever will. The group is strongly 'closed' at this stage and it is very difficult for new members to break in or to be introduced.

Development from one step to the next cannot be forced: group cohesiveness seems to depend on how well the members can relate in the same phase at the same time. The group will proceed through these five stages at the speed that its members are prepared to accept because each member must be prepared to give up something at each step in order to make the move to the next; to grow from the polite step to the 'why we are here' step means that each member must relinquish the comfort of non-threatening topics and risk the possibility of conflict. The move to the 'bid for power' step means that continued discussion of the group's purpose must be put aside and individuals must commit themselves to one with which they may not feel completely at ease. Growing to the constructive step requires individuals to stop defending their own views and to risk the possibility of being wrong and, consequently, some humility is demanded of group members at this stage. The final step from phase four to phase five demands that a member trusts himself and other group members, the danger here being that to trust is to risk a breach of trust.

8.5 Effecting change

Resistance to change will be a day-to-day experience in the life of a systems project leader. Sometimes this resistance will come from the client organization, for whom the project is being carried out, and sometimes it will come from team members who will resist the changes being implemented by the project manager in the management of his team. Good project leaders, therefore, will have planned a number of approaches to remove blockages to change and to achieve project success. A general model – known as force field analysis – is useful in this context.

Force field analysis is a tool for analysing a situation that you want to change and is based on the fact that, in general, the present situation is a result of a balance between forces (Figure 8.6).

When it can be seen more clearly what these various forces mean in practice, there is a better chance of bringing about change in the desired direction. Once the forces on both sides have been identified, they can

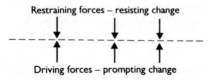

Fig. 8.6 Force field analysis

be assessed: some will be strong and significant, others will be of little consequence.

Change can be brought about in two ways: either by increasing the forces prompting it or by reducing the forces resisting it. By adding them to the driving forces and weakening the resisting forces, there is a much better chance of achieving constructive change. If, however, driving forces that threaten or pressure people are increased, resistance (or resentment) is likely to increase. It is normally better, therefore, to increase those driving forces that do not increase resistance, and to work at reducing the restraining forces, or to consider new driving forces that may be brought into play.

Within the client organization, senior management should have been persuaded to let employees participate in the development of the new system so that they do not see the project team as an external catalyst of change to be resisted. The project leader will be aware of their fears of loss of job, status, security or colleagues, and listen to them sympathetically, appreciating their contribution and helping them to develop into their changed roles. People often cope with change better if the ideas and plans are introduced to them sufficiently early so that there is time for the effects and benefits to be fully explained and understood. The longer the time between announcing and effecting the change, the lower will be the resistance to it. However, the longer it takes for benefits to be realized, then the longer will resistance continue. Therefore, within the team there is the need to build a sense of maturity which is sufficient to cope with the pressures of uncertainty and the lack of instant achievement.

Participation in decision-making should reduce resistance to change in proportion roughly to the amount of participation. Flexibility in approach from the project leader will also reduce resistance and, when possible, transition from the *status quo* to the new conditions should begin only when a consensus has been reached about how it is to be effected. If flexibility and even change are a part of a team's way of life, members will be constantly open to new ideas and methods and prepared to accept changes that are properly planned and communicated, and will be fairly evaluated at the end of the day.

8.6 Rekindling motivation

The team may achieve a certain amount because it is motivated to do so, or under pressure to do so, but if the operational climate is not healthy, then in time members will become dissatisfied and consequently de-motivated. In these cases, the project leader needs to examine his judgement of the situation and see whether there is a gap between his observation of the team's feelings and the reality of the situation.

Group meetings in these times of dissatisfaction tend to be chaotic, or may even promote a hostile situation, and it is wise therefore to speak to as many individuals as possible in order to understand the general climate of the group. Once this climate has been evaluated, the project leader can hold a meeting with the whole group, fully prepared to understand and to compromise where possible, and to solve problems.

If the whole team is under-achieving, then there is likely to be something wrong with the operational climate; but if just one or two individuals are lagging then the project leader can try to enhance their motivation by using some of the practical methods that have been explained earlier in this chapter. If he has to deal with the under-performance of a specific individual who lacks motivation, he needs to assess whether this results from an inability to perform the task, an inability to work in the team or a lack of supervision.

An inability to perform a task may be rectified by providing a piece of formal training but, if this is not the case, then the project leader has the problem of removing them from the project. If they lack the ability to get on in the team and are even counter-productive, then they can be counselled, and it should be demonstrated to them how their behaviour appears to others so they can have the opportunity to improve. In the extreme, however, they may need to be taken off the project too. If people are de-motivated through inadequate supervision, then it falls to the project leader to put this right. In each of these cases, it is worth him referring back to the selection process he went through to see if he can identify errors which led him to select people who it now proves are inadequate to the task.

Only when methods of re-motivation have failed should the project leader resort to coercion methods. It should be recognized, however, that there will be situations in which this will become necessary because allowing unacceptable performance or behaviour is not an option which can be allowed to run for very long if the project is to succeed.

8.7 Summary

Each time a project leader is appointed, the requirements of the job will be different. In time, experience will be gained of dealing with different people, in different situations and on different projects, but the same basic ground rules apply for each project. The project leader must understand his own personal needs, which vary over time, and must get to know all his team members and what they need to get from the project too. Practical measures should be taken to create a motivational environment and this should be maintained through his daily management style, behaviour and actions. He should try to develop his team's effectiveness and maintain motivation at times of stress, and be quick to take remedial action when it is necessary. His abilities in these essential project management tasks will contribute substantially to the success or failure of the project.

9 Performance improvement

9.1 Introduction

Defining and agreeing the performance that a project leader expects and needs from each individual is an essential requirement for the successful management of the project. He will, of course, have communicated the overall objectives, gained commitment to these from team members and allocated to each the specific tasks for which they are responsible. At the start of the project, therefore, every member of the team will be aware of the tasks, the standards, the quality and the time-scales to which the project leader is working. They will also have been aware that these are likely to change over time and that the development of a mature response to this kind of uncertainty is the sign of a well-organized team. As part of the motivation plans, the project leader needs to make every team member aware that they will have adequate feedback, support and development, as well as the appropriate resources and job environment for them to achieve satisfactory performance. The concept of a 'contract' between the two parties can be very helpful even if most of it is unwritten, and in planning his objective-setting discussions it is useful to list what he expects from the members of the team and what they are entitled to expect from him and the organization. This will no doubt be revised after the discussions and the consequent record will be a useful guide to motivational needs, as well as adding clarity and focus to the direction of the efforts of the individual.

9.2 Objective setting and performance standards

An objective is a statement of a desired result to be achieved by a set date. Therefore, the project leader needs to set team objectives accordingly and monitor actual performance against these. This is an essential part of the project planning and control process and a pre-requisite to assessing and subsequently improving the performance of the team. Team members must work towards the achievement of tasks that are compatible with one another and possible in the light of the resources available. Co-ordination is needed to avoid differences of opinion about how objectives should be achieved and even about the objectives themselves and, while co-ordination planning is time-consuming, it is a wise investment. It is in many ways analogous to that of getting to know the team staff at the start of the project, since it is much more efficient than trying to tackle a co-ordination problem once it has occurred. The project leader should involve all the team in setting the objectives at an initial meeting; some experienced members will add to his knowledge and those unable to contribute will nevertheless feel excited at and be developed by being involved in the planning stage and watching the project grow. When every member comprehends the project's overall objectives and their own personal ones, they are more likely to be motivated to achieve them. In doing this, the project leader must try to engender a genuine feeling by all team members of 'ownership' of the team objectives.

It is useful to think of a 'cascade' in the building of a hierarchy of objectives that, when fitted together, produce the desired result. The project leader should start at the top by describing the overall purpose of the objective: in one or two paragraphs state what purpose the successful achievement of the project will fulfil (e.g. to produce management information for the managing director faster, cheaper and efficiently). Then he could list the aims of the project itself that, when completed, will result in the purpose being achieved. Each aim can be shown to be able to be achieved by setting quantified objectives for teams and then individuals. The number of levels in the 'cascade' will depend on the size and complexity of the project, the number of teams and the number of management levels. We have seen in a previous chapter how involving people in setting the objectives and giving them responsibility for achieving their own tasks are sources of motivation in themselves. Every team member therefore needs to know whether their individual objectives are being achieved and how the project as a whole is progressing. Every member should have a task description that lists the general tasks and standards to be achieved, and a 'guide' that lists team objectives should be given to each individual. These underpin the whole objective-setting process and enable motivation to be built up.

The objectives guide is divided into two parts: the first part lists important areas of the job in terms of the results needed; the second part lists the special tasks on which the team member is working, together with the date by which it has been agreed they will be fulfilled. Of course, people have greater commitment to objectives if they themselves are allowed to analyse the job, albeit with the guidance of their project leader. They can list their own objectives, which will enable them to achieve their goals, and agree with their project leader how they will achieve them, so that both can discuss opportunities for improvement and potential difficulties. Less experienced team members can be given objectives and allowed to work out how they will achieve them.

The project leader can therefore give every team member a statement of the overall objectives of the project and how these will be achieved *after* their mutual meeting, so that he can clearly show that he has taken account of the suggestions and opinions of team members and that he is expressing a consensus plan of action and priorities. As the project progresses and shortfalls in performance and changes to plans have to be tackled, this early consultative approach will really pay off.

Objective setting gives the following benefits to the individual:

(1) They know what they must do.
(2) They have a yardstick by which to measure themselves.
(3) They can organize themselves to meet the objective(s).
(4) They have an opportunity to accept greater responsibility.
(5) They have an opportunity to express their strengths and weaknesses and state areas in which they feel uncertain.
(6) They have an opportunity for development and recognition.

To the project leader, objective setting is useful because it:

(1) Concentrates on priorities.
(2) Leads to improved information and controls.
(3) Improves morale.
(4) Identifies the contribution of each team member.
(5) Identifies individual training and development needs.

The essential characteristics of a good set of objectives are that they are:

(1) Relevant.
(2) Within the control of the individual.
(3) Realistic and achievable, but stretching.
(4) Specific, clear and precise, not too general.
(5) Flexible in how they are achieved.
(6) Have a time-scale.
(7) Are not too many.
(8) Are consistent with other people's objectives.

When designing objectives, therefore, the project leader must promote team work and not competition. He must provide for goal-sharing where possible, create an open climate and avoid building feudal boundaries.

Goals should motivate the team more than control it: people can be inspired to greater efforts by the challenge of their jobs and, therefore, objectives should be designed to increase the demands on their abilities so that they can achieve some personal fulfilment of their potential in the achievement of the goal.

9.2.1. Performance review

There is no point in setting goals unless success in achieving them is reviewed. Team members will automatically take note of their own performance and, if motivated to do so, will take steps to improve or correct it.

Informal get-togethers should therefore be a way of life for the project team to reinforce this process. The project leader needs to ensure that there are sufficient informal meetings for everyone to know what is going on, to receive feedback on what has been achieved to date and to help plan for any necessary changes in his courses of action. If he is an open, sympathetic and accessible project leader, people will come to him informally with their work problems; and then he should gain enough sensitivity to know when to approach people about their difficulties. All the time, he needs to encourage people to spot problems, cope with difficulties and find their own solutions so that he promotes confidence and independence. In the end, however, he is responsible for reviewing performance against plans and objectives, and for taking corrective action and getting others to do so. In doing that, it is hoped that performance reviews are seen as re-planning sessions rather than post-mortems.

The flowchart in Figure 9.1 is a useful guide for carrying out these kinds of reviews and continually improving on the targets set, to match revised needs and time-tune the management system.

A project leader should start in the box 'Are the planned results being achieved?'. He should be sure he knows the facts; he shouldn't go on his own or others' perceptions alone. If not reasonably sure, then he goes down the 'Don't know' route and improves his knowledge, information methods or clarity and communication of objective. If he knows there is a shortfall or a changed requirement, the 'No' route gives him five

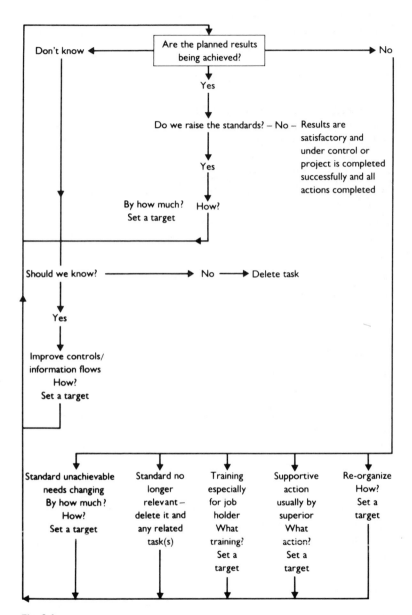

Fig. 9.1

possible reasons and subsequent actions. As you will note this is a cyclical process; more discussion on the actions to be taken follows.

9.3 Variations in performance

If the project is sufficiently well planned, the staff sufficiently well selected and motivated, most performance shortfalls, where they occur, will be as a result of influences external to the team. However, if all is not going according to plan, the project leader needs to instigate remedial action. Figure 9.1 aids the diagnosis. It may be that some small adjustments are all that is necessary; or there may be a need to undertake significant re-planning or re-directing: perhaps changes or circumstances that could not be foreseen have arisen and are causing performance to vary. If the team as a whole is under-achieving, the projet leader must ask himself:

● In what areas is it under-achieving?
● To what extent?
● Was the plan unrealistic?
● How can the plan be altered, and alterations acceptable to the client made?
● How can performance be improved?
● How can he communicate the performance shortfalls without de-motivating the team?

If he has engendered good team spirit and if he is able to treat problems calmly and constructively, the project leader might well be able to express the need for remedial action as a challenge which the team can enjoy and from which it can learn and grow stronger.

In political or military circumstances, very often a good speaker can generate a willingness to offer exceptional commitment and, in difficult project times, the project leader must influence his team to want to work their way out of problems. At a well-prepared meeting, he must make clear that they are all 'in it together' and that in no way is he judging their and his own performance, but merely stating facts and looking forward.

If the project leader has developed an 'ownership' of the team objectives by all the members, this process is more likely to be positive. He needs to ask team members for opinions and suggestions, and together form a plan of action. At the same time, he must ensure that strong, positive, internal team feelings do not turn into negative feelings towards those outside the team; this is a common reason for once-successful teams failing. Where he is dealing with individuals who under-achieve, he must ask himself:

● If he expresses their performance shortfall to them, are they *able* to improve?
● How to communicate the need to improve without alienating or de-motivating them?

It may seem obvious to say that the project leader should not point out the need for remedial action where this will have no effect. In stressful circumstances, however, in which his own position could be threatened, he might find himself wishing to castigate team members because of his need to 'get it off his chest' rather than for the sake of the project. At these times, the understanding of his personal needs and feelings and the constant referring back to them will underpin all his tasks as a leader and manager; this was discussed at the beginning of the previous chapter. Where remedial action can improve performance, he can start the improvement process for an individual by having a discussion using the process outlined in Section 9.5 (i.e. preparation, establish and agree the performance gap, explore reasons for the performance gap, agree steps to eliminate the performance gap, summarize, follow up).

9.4 Appraising performance and identifying training needs

Whether or not the project is progressing to plan, people like to know how they are performing, and so the project leader should build into his original, overall plan an appraisal system that will enable him to review performance at useful intervals during the project. Naturally, when there is a need to improve performance, he will have to bring forward dates for appraisals in order to identify immediately the problem areas and the consequent training needs.

Of course, informal appraisals go on all the time, for instance, 'I think he can do the job better than him' and 'If I'd done this differently, he would have done that better'. However, sympathetic information regularly gathered and reviewed with an employee on a more formal basis, annually or at the end of projects, is essential for the following reasons:

(1) It is as fair and objective as possible and seen to be so if properly executed.

(2) It assures team members of an opportunity to discuss their performance and problems with the project leader.

(3) It enables the project leader to raise matters privately with individuals, having had input from team members to advance his understanding.

(4) It gives a synthetic view of all the human resources available during the project on which the project leader can draw.

(5) Team members are motivated to work for the project leader because they perceive him to be interested in them and because he seems to have the use of human resources under control.

In this appraisal process, the needs of the individual team member and the project itself are complementary. The individual will need to know:

(1) *What's expected of him.* This is clearly impossible to handle properly unless some objectives have been set and agreed beforehand.

(2) *The project leader's view of him.* The criteria for forming a judgement need to be open and understood by both parties and applicable to the job that is being done.

(3) *Other team members' views of him.* This will need to be handled carefully to avoid setting individuals against each other; it is important, however.

(4) *Client relationships.* He should be careful here to stick to facts and those that are relevant to the objectives agreed.

(5) *Technical performance.* This should be the easiest to quantify. People either have a grasp of technicalities or they don't.

(6) *Individual output,* as measured against the agreed objectives, i.e. an individual's output, not their input. Also, he must concentrate on achievement of objectives, tasks and standards, and not the personality characteristics he has observed or the motivations he suspects.

(7) *Team performance.* Giving feedback on contribution to team performance can be very rewarding.

(8) *Personal contribution to the project,* e.g. loyalty, dedication, willingness to accept personal inconvenience: the inputs the individual has offered.

(9) *Management skill/potential (if applicable).* This will range from ability to help lead and train others on the job to promotion to larger management positions.

(10) *Own training needs and learning opportunities.* Recognize what has been learned in the past before talking about the future. Match training and learning needs to project personal success.

(11) *Opportunities for development,* so that personal potential can be realized.

If the project leader is carrying out appraisals for the first time, it could be helpful to enlist the help of someone with some skill and experience in appraisal and counselling; perhaps his company or client have an appraisal system that can be used. Certainly, he will need a set of forms that will cater for the procedural needs of the appraisal process and as a record of the outcome of discussions. An example of the sort of procedure and process recommended is shown at the end of this chapter.

The appraisal interview should be carried out with the same tact, sensitivity and objectivity required for a selection interview. The project leader should ensure that he has time to do all the interviews so that he gains a total picture of the project and assesses everyone by the same standards.

He should see his role of appraiser as being helpful rather than judgemental and make certain that there is a two-way conversation between the appraisee and himself so that the team member being appraised is as free as possible to comment.

In fact, the more the individual is able to appraise their own performance and describe it the better. Both parties must be properly prepared for the interview, the appraiser and appraisee having read each other's comments on the form beforehand. The purpose of the appraisal is to record performance to date and to set objectives against which it can be measured during the next stage of the project. Both parties need to participate in this wholeheartedly.

Where a team member is under-achieving, the project leader should look at the original tasks that were set and see if they were unrealistically demanding or if he had set priorities that could not be achieved. However well or badly the appraisee is performing, the project leader should seek to measure the quantity and quality of his work and the time taken to complete it. The project leader should also underline any specific tasks where improvement is instrumental to the project's success.

The simple knowledge that targets have been achieved or not is not sufficient for appraisal purposes. It is equally important to know more about the effort and the skills that were applied in the light of the realities that developed during the period.

How strong was the team member's commitment and how competent were his or her attempts to improve matters? How good was judgement? And so forth.

One safeguard in reaching a realistic view of all the team lies in the design and structure of the appraisal forms. An identical form and procedure for all staff assures that everyone is assessed in the same way. The leader or manager must avoid marking too highly because he is lenient, or too harshly because his own standards are very high. The basic requirements of good appraisals are as follows:

(1) They should explain the purpose and methods of the appraisal so that appraiser and appraisee are clear and in tune with each other.

(2) They should demand specific and not just general information. Space should be given for descriptions and not just marks or grades.

(3) They should make it necessary to formulate action plans, such as the achievement of objectives that may include training.

(4) They should require signatures from appraiser and appraisee to show that both have read and agreed the description of the past, the discussion about it and planned action for the future.

So far as training is concerned, the role of project leader is first and foremost to fulfil the project's requirements. There will be cases where training is vital in order for someone to achieve their project tasks; perhaps because these tasks have changed over time or because the individual joined the team for reasons other than complete suitability. In such cases, the project leader must gain people's commitment to training by explaining that it increases work opportunities, job interest, involvement and so on. Often, people are very happy to be developed when training is agreed with them as an opportunity for growth and not told to have it as a necessity for overcoming their deficiencies.

The project leader may be able to arrange on-the-job training for some technical skills, while the development of inter-personal skills may require attendance on a specific course. He should always discuss with his people what they have learned, so that they consolidate their training and he can plan how best to assist them to implement their new skills. After attending a course, individuals should complete a training evaluation form: a typical example is shown in Figure 9.2. The project leader may find it helpful to add comments to this form, since they will serve as a useful record and guide for the future. He should observe whether their new knowledge or skills are being applied, so that he can work out the cost-effectiveness of the training.

9.5 Improving performance

While the use of the appraisal system and of training can improve performance, there may be instances where individuals fail to fulfil objectives agreed at appraisal time or to respond to their training.

The following performance-improvement process can be used

Training evaluation form

Name _____ Job title _____

Location of course _____ Company _____

Course attended _____

Name of organiser (external course only) _____

Date attended _____

1 What was your overall assessment of the course?

   Excellent Good Satisfactory Fair Poor
       5   4   3   2   1

2 Did the course achieve its objective? If not, please state reasons

3 What were the most important points you learned from the course? How do you intend to apply them to your job?

4 Which part of the course was least useful and why?

5 Any recommendations you would make to improve the course in terms of content, method of instruction, balance of lecture, roleplays etc?

6 Any comments on the location and arrangements for meals etc?

7 Comments on the instructor(s) taking into account the following: knowledge of material, method and style of presentation, responsiveness to class, strengths and weaknesses

Signature of trainee _____ Date _____

Signature of manager _____ Date _____

Fig. 9.2 Training evaluation form

informally in a short performance-improvement conversation, semi-formally in a performance-improvement meeting, or formally in a disciplinary interview, when all informal methods have failed. The steps to be followed in each circumstance are the same and are:

Preparation
The project leader must ensure that he investigates all relevant issues so that he has all the relevant facts – not opinions – at his fingertips. If he is feeling angry, let down or negative in other ways, he should wait and have the discussion later. He must separate behaviour and its effects from perceptions and interpretations. If it is a formal disciplinary

meeting, he should interview in private and follow the company procedure. It is imperative for him to have a clear objective for the meeting and to know what he expects to get from it and what he wants the individual to do differently in the future.

Establish and agree the performance gap

At this stage, the project leader should try to avoid recriminations and not get sucked into discussing the reasons for the problem or the mitigating factors that may apply.

He should be positive, logical, rational, objective and non-threatening in his attitude, and firm, managerial, sensitive and equable in his behaviour. As in any interview, he should use open questions and keep summarizing what the person has said to confirm this with them.

It is imperative that he listen carefully to all that they say so that they feel they have been given a fair hearing and that he fully understands the situation from all viewpoints.

He should be sure of his standards and expectations and ascertain whether these are understood by the individual. He must achieve agreement on expected performance in order to establish the extent, measure and effect of their actual performance and the degree to which these vary from that expected. He can adjourn where necessary; for instance, to re-assess the situation. Only when a clear understanding of the problem has been reached should discussion move on to the next section.

Explore reasons for the performance gap

The project leader must investigate whether the shortfall in performance is a disciplinary matter or whether it results from one or more of the following:

- poor job description;
- inadequate workplans;
- lack of understanding of expectations, standards or rules;
- poor selection or training;
- medical or domestic problems;
- a genuine grievance; that is a request or objective was unreasonable, unachievable or illegal.

Now the project leader is at the stage where he can ask for reasons, mitigations and so forth, provided it is a disciplinary matter and he has *sufficient facts*, and on the condition that either agreement on the gap between actual and expected performance has been reached or that disagreements are well understood. Also, there should have been adequate discussion with the personnel department and/or the individual's manager for him to continue: if not, he must adjourn in order to investigate further.

Agree steps to eliminate the performance gap

What can the project leader do? He can remind the team member of the expected performance and agree an improvement plan with positive changes that will occur in specific time periods. He should write down all this and, if a formal process, should confirm it by letter later. He should express encouragement and offer specific help where he can, bearing in mind that through this procedure he is seeking to improve performance, not to worsen it by de-motivating the person.

Summarize

Specific points must be summarized in discussion with the team member to make it clear that this is a disciplinary procedure, albeit informal.

The project leader should explain the expected and actual performances and the reasons for the gap between them; he should underline which reasons are acceptable and which are unacceptable. He should proceed to summarize in detail the actions, commitments and agreements for change and on what date performance will be reviewed, stating what will happen if it continues to be unacceptable. If this is a formal meeting, the discipline should be given with a sanction and all the points brought up should be given in a letter to the team member and a copy of this kept on their personal file.

Follow up

This should be done on or near the date specified. The project leader must start at performance and continue the process until either performance is totally acceptable or the person leaves. If performance is now acceptable, he should write a letter of praise to promote motivation.

The process of going through the above model with an individual can be started at the most informal, conversational level and can be made slightly more formal each time with more rigorous and well-defined outcomes for action and greater sanctions for non-performance until, finally, dismissal can occur.

The disciplinary interview is not a contest—no one wins and no one loses. It is an instrument for getting performance or behaviour back to the required standards. Discipline at work is not about punishment; it is a management tool which needs very careful handling, for breaches of discipline can lead to undesirable precedents, bitterness, de-motivation and accusations of favouritism and unfairness, and to overlook them is to store up more problems in the long run.

On occasions, a shortfall in performance may be the result of a grievance or grievances which an individual or a group is harbouring, in which case they need to be resolved. The following checklist can be used to set the framework for handling grievances:

(1) Are the team members aware that they must raise all grievances with the project leader in the first place? If not, this should be covered at selection or the start of a project.

(2) Do they understand the procedure? Does the project leader? He may wish to enlist the services of a personnel department, but typically the steps are first to raise the issue with the immediate superior with the right of access to the next (or a senior) level of management if unsatisfied.

(3) Is it a proper grievance? For example, a complaint against a fellow employee's behaviour is not a proper grievance although it should not be ignored. Examples of grievances that should be raised are where an individual is penalized for non-performance of an objective of which he was unaware or where achievement would have been almost impossible. Also, for example, where an individual was promised a pay review or job responsibilities that did not materialize.

(4) Has the project leader sufficient information to answer the grievance? Does he need to refer the matter to his superior? He must try to avoid unreasonable delay and should say when he thinks he will be able to reply. It is important before doing this, however, to find out if there is an appropriate rule, procedure or bulletin to which he can refer the employee.

(5) Is a witness needed? Circumstances when it is necessary to have a witness are where the individual has made it clear it is a formal grievance, where the issue is quite a serious one or where a lot of 'opinion' is around and a witness to what was said is necessary.

(6) Is the project leader sure he has all the facts and the opinions? Has he reported them fully to his boss if required to do so?

(7) The project leader should try to put himself in the complainant's shoes. He will then avoid some of the more obvious pitfalls, ensure he has gathered all the facts and, having seen the issue from both sides, will be able to be more objective.

(8) Once the grievance is settled, the project leader should check with the individual or group afterwards to regain confidence and re-establish effective relationships and teamwork.

Handling a grievance

First, one must be sure that a genuine grievance exists. Here is an example of one and how one might go about it, followed by the general approach to be taken. If an individual is selected by someone other than the project leader, who joins his team and after a few months claims that he was promised some supervisory responsibility, the project leader would need to check the offer letter, talk to the recruiters, check the actual words used, counsel the individual on his real wishes, and clarify what he thinks was actually said. If, say, supervision was suggested as a possibility after project success, the project leader now needs to reach a new understanding with the individual, confirm it in writing and follow it up after a period.

Before the interview, the project leader should examine the issue, gathering all relevant information and making discreet investigations where necessary. Once he has all available evidence, he may begin to assess the facts. He should then decide upon the objectives and/or actions in dealing with the problem and should consult the personnel department if in doubt about the action he can take. He should consider what compromise could be made with the team member. He must prepare notes on all the factors for his reference on a grievance case and read through them before an interview with the team member.

During the interview

It is essential that adequate time is allowed for the interview. The project leader must plan against interruption, e.g. switch off the telephone. As in all interview situations, he must put the subject at ease. He should then listen without interrupting while the complainant states the case and show interest in what is said. He must then state his case and reply concisely and precisely. Discussions should be kept cool, and he must not get drawn into an argument.

The project leader should summarize what the complainant says to show that he has listened and give a chance for corrections. Also, this provides an opportunity for the project leader to draw them out where necessary and helps them to see their problems in perspective.

If in doubt he should adjourn, e.g. he needs more facts, he needs to calm them or himself down, he may need to check out possible options or refer to higher authority.

If agreement is reached, he should inform the complainant of the next step to be taken. He should always follow up the results of the interview and make sure that the entire grievance procedure and its results are recorded on file.

After the interview
The project leader should act upon any agreement he has reached with the team member. If the matter is not settled, he must report the grievance to his supervisor.

9.6 Help from outside the team

A variety of people outside the team also contribute to improving its performance through the way they contribute to its activities.

The project leader should establish good relations with departments that supply office space, manpower, accounting and other information etc. and get them involved in the project so that they understand what is done with the resources they provide and how together they can work towards a successful implementation and the overall success of the organization to which they both belong. If, for example, he treats administrative services as an obstacle, they will behave like one.

If the project leader ignores their feelings and just expects them to do the job they are paid to do, then he will get grudging low performance from them (of course, this is true of people in his team as well).

One of the most essential preparatory exercises is to think through who is affected by any changes and recognize that individual's feelings about change need drawing out and not smoothing over. Even what to the project leader might appear a positive benefit from change will not necessarily be seen that way by users of the system. He can list the positive forces assisting him achieve his objectives and the negative ones restraining him.

Research has shown that working on removing the restraining forces (or obstacles) is much more productive than trying to increase the positive forces as this often only increases the negative ones.

9.7 Summary

Good system design, together with efficient and sensitive use of human resources, should mean that the level of performance on the project is acceptable. In fact, successful performance and recognition of it promote even better performance and, if the team is well-selected and well-motivated, it should be able to withstand pressure, changes in plans, alterations to schedules and so on.

When there is a short-fall in performance, the project manager needs to investigate the reasons for it and take the appropriate actions, always conscious of the effects of his or her behaviour on the team. His skill as project leader lies in his ability to deal with such issues of change management in a positive and constructive manner.

9.8 A performance appraisal system

If you have used a formal appraisal system in your work before, you will be familiar with the methods and procedures as well as the principles and philosophy behind it and the various ways in which it can be used. An appraisal system has many objectives and opens the door to many opportunities. Some of these are described in the following appraisal system which is used in a large UK computer services company. It is couched in formal language since it is taken from a real operational environment.

Performance Appraisal System

1. *Introduction*
The company's policies provide for the assessment and recognition of performance; the assistance with continuous personal development through training, new experiences and other methods; the identification of employees with potential to develop further; the setting of clear aims, objectives, tasks, methods and approaches.

2. *Objectives*
The prime objective of the appraisal is to help ensure the maximum contribution of each employee to the corporate aims. Assessment plays its part by

(a) Assisting in conserving and improving employee skills and capabilities through the careful identification of individual abilities and potential and by motivating employees to give of their best.

(b) Recognition and recording of achievements, successes and strengths.

(c) Helping individuals to improve their performance in their current job.

(d) Setting new targets for each individual to further their own development and the achievement of corporate aims.

(e) Identifying the immediate and continuing development and training needs of each person.

(f) Encouraging self-development.

At the same time, the assessment procedure will help employees to answer those questions which they frequently ask themselves:

How am I doing?
What can I do in my current role to prepare myself and improve and develop for the future?
Where do I go from here?

3. *Procedure*—For completion of Appraisal Form A (Figure 9.3)

3.1 Appraisal Form A is concerned with a review of performance to date. The two people concerned with the appraisal, the appraiser and the appraisee, are both sent Form A to complete in draft. The overall aims and objectives will be filled in by the appraiser unless they are taken from the previous appraisal's workplans. Form A is sent to the appraiser and the appraisee about a month before the appraisal is due and the appraisee also receives a note showing his personal details and the training record which has been kept for him so that this can be checked and updated where necessary.

3.2 An appropriate date and time is organized for the appraisal discussion.

3.3 At the beginning of the appraisal, or perhaps beforehand, the appraiser and the appraisee exchange their own versions of the completed form as preparation for, or as the basis of, a discussion during the appraisal interview.

3.4 During the appraisal discussion, achievements, successes and strengths are recognized and differences in views are discussed and agreements reached as far as possible.

3.5 Eventually, the appraiser and the appraisee both sign the form to show that it is a record of the performance over the last appraisal period and then the appraiser sends this to his manager to review, agree and sign, so that it can be recorded in the appraisee's personal file which is held by the personnel department.

3.6 A blank form is used to document the aims, the objectives and the other workplans for the next twelve months and this is then used as the basis for the appraisal next time. The appraiser will then complete the personal details on the new form, transcribing them from the old one.

3.7 Achievements may also be reviewed on an informal basis from time to time and the relevant review and revision information is documented in the appropriate columns on Form A under the heading 'Objectives Review'.

Appraisal/work planning form   *Strictly confidential*   *Form A*

Personal details

Name _____ Job title _____ Department _____

Site _____ Time in present pos. _____ Age _____ Service _____ Date of last appraisal _____ Papers issued _____

Appraisal period: From _____ To _____ Show position in a 3-level hierarchy (note total numbers managed where applicable):

Training and qualifications since last appraisal

_____

_____

_____

Cost budgets   £ _____

Revenue targets   £ _____

Contribution targets £ _____

Purpose of job:

Overall Aims: | Objectives and other work plans in the appraisal period (group under the Aims Headings
(No. 1, 2, 3, etc.) | 1, 2, 3, etc. and continue over)
1. | 1. (a)

Fig. 9.3

Objectives review

Obj's (as per above)	Reviews and revisions (continue over)			Annual (what was achieved, how; what was not achieved, why?)
	1st	2nd	3rd	
1. (a) etc.				

Fig. 9.3 – (contd.)

Overall Review

1. *Overall achievement of objectives and standards including relationships with others*

2. *What has been done best during the year? Main contribution to the business*
   What has helped this? How can this be repeated and built upon?

3. *Which areas have shown least success?*
   What factors affected this? How could improvement be achieved?

4. *Which changes would most significantly influence achievements of objectives over the next year?*
   Who should do what, when and how?

5. *What other changes should occur (e.g. by individual, manager, company)*

6. *What are the strongest areas of personal skill, what personal learning and development has occurred?*
   How might these be developed and utilised over the year?

7. *Which areas of skill, knowledge or experience could be helped most?*

8. *What specific training or experience is recommended? (Or education, projects, coaching, secondments etc.)*

9. *What do you see over the next 3 years?*
   Likely time present role, what when, which career direction/options? New skills to develop? Types of job to develop into? Actions to assist the directions? Which sorts of jobs should the appraisee be considered for when they arise?

10. *Any other comments, questions, enlargements, etc.*

11. *Comments by appraisee*

Appraisee _____   Appraiser _____

Date _____   Date _____

Comments by Appraisers Manager

Fig. 9.3 – *(contd.)*

3.8 Employees receive a full appraisal six months after starting a new job and every twelve months thereafter. When an individual changes job, the losing manager carries out the appraisal or the quarterly review, if there has not been one within the last six months, and a new Form A with revised aims and objectives is created by the new manager so that the next appraisal date is then six months from the date of job change and then annually thereafter.

10 Managing quality

10.1 Introduction

Before discussing methods of improving the quality of the systems planned to be delivered, consideration should be given to what is meant by quality. It will help to think in terms of the quality of a consumer-durable such as a washing machine, a suit of clothes or a piece of furniture, and relate requirements for it to what is expected from a new system. Requirements would probably be that the product should work properly and be reliable, be well made, last a long time, look good, be easy to use and feel comfortable in use. It may have cost a little more than a poorer-quality product, but one should be happy to pay the difference because of the utility of the product. Also, we would want to be proud of the product and pleased to show it off to our friends.

However, a good-quality product may not be particularly fashionable or have all the very latest features; it is more important that the product meets the desired fundamental needs.

If we translate our analogy from a consumer durable to a computer-based system, we can examine quality in more detail. For a computer-based system to work properly, it must be free from errors and not prone to breakdowns, even when the system is put under greater pressure than originally intended. Clearly, then, thorough testing of a computer system is important to eliminate faults, ensure resilience and establish its reliability under operational conditions. The system must also meet the needs of the user; it must be complete and comprehensive and do what the user wants. The delivered system must fulfil the original requirements specification. Here, the testing of the system should not only be to eliminate faults but to prove the design.

A good-quality product that has been well made also uses good materials. However, a system that meets its requirements and is free from errors does not necessarily mean it has been well made. In the context of a computer-based system, a well-made system is one that is properly structured and is, in consequence, easy to maintain and modify. The component parts will fit together according to a logical plan and not be haphazardly assembled. The materials employed must also be of good quality. The program code must be easy to read and there should be no skimping of instructions dealing with exceptional routines. Adequate messages must be incorporated into the body of the code so that it is readily understood by others. The logic employed must be simple and not convoluted. Systems designers and programmers must not include intellectual niceties for the sake of it.

The system needs to be designed and built to take due regard of its use of machine resources. That is not to say that it must minimize the use of memory, storage and computational power, but that it must not use these resources excessively. Good design and coding should give an efficient system with lower running costs. However, with the continuing drop in memory and machine costs, a balance must be struck between efficiency in the use of these resources and the ease of development of the system. A good-quality system is unlikely to use machine resources inefficiently but when the choice is between thoroughness and simplicity in design, and machine efficiency, then the former must prevail. The significance of machine efficiency will depend on the installation standards for the recovery of machine costs or the machine resources available for the project. A system being developed for a microcomputer configuration, for example, will impose particular constraints.

A system that has been designed to meet the user's requirements is easy to maintain and enhance, not excessive in its running costs and will last a long time. This is true even when new technologies and techniques evolve which supersede those used in the original system. The high cost of development of new systems will ensure that the old, good-quality system has a long life.

The system will be easy to use if care has been taken in the design to take advantage of the experience and expertise of the users and of the circumstances and environment under which it will be employed. It is not good practice, for example, to assume that the users understand how the computer and software utilities work! Good project development procedures and methods such as prototyping help to ensure that systems are developed in line with the user's capabilities. If the system has been developed so that it does not demand too much attention or knowledge from its users, does what is expected of it and behaves in a sensible and reliable manner, then they will feel comfortable with it.

Finally, if the system has these qualities, one will indeed be very proud of it and, like the good-quality consumer product, will be very happy to show it off to friends and associates.

A good-quality product is often more expensive than an inferior one. The higher price should be weighed against the better points of the product to justify the choice. The same argument can apply to the development of a good-quality system: it may well cost more to develop but the payback is a longer life and lower support costs. This is discussed in more detail later.

To summarize then, a good-quality computer system is

Error-free
Complete
Comprehensive
Reliable
Easy to maintain and enhance
Flexible
Efficient

A word of warning, however: the quality of the computer system must not be confused with the quality or appearance of the documentation or of the hardware. They are a physical manifestation of the application system and they also need to be of good quality, but they should reflect the quality of the application system rather than attempt to be a substitute for that quality.

Furthermore, a good-quality system, by the nature of its design, can minimize the need for exhaustive documentation.

10.2 Importance of quality

If a system is of such poor quality that it does not meet the user's requirements and contains errors, then the user will be dissatisfied.

Depending on the contractual arrangements or on the method of funding maintenance, the user may be in a position to insist on the corrective measures being made free of charge to him. This could have disastrous financial implications to the development organization.

Even if this is not the case and maintenance is regarded as an overhead as it is in some organizations, there will be ill-will and a lack of confidence in the development team. This will create problems for any subsequent projects; indeed, there may not be any.

On the other hand, if a good-quality product has been delivered, then the client or user will be satisfied. This satisfaction will increase with time as he learns how reliable the system is and how it can be modified to meet his changing requirements. This will create demand for more systems of the same quality. The development team will therefore spend their time building new systems instead of maintaining old ones, and business efficiency will be improved.

A project leader will often be working under many pressures such as a shortage of resources – both human and machine – delivery deadlines, development budgets and so on. You may think that much of what has been said so far about managing quality is all very well, but that it is practically impossible and will cost too must to develop. It must be admitted that implementing the steps necessary to ensure a good-quality system could incur greater development costs and delay delivery dates. However, with experience and practice, these steps will become second nature and a normal part of the development and should not impact costs and delivery dates. In the final event, even if it does cost more to develop a high-quality product, longer-term benefits of this must be considered, such as reduced maintenance costs and increased user satisfaction.

A computer-based application system can have a life-cycle of typically seven years; many systems have a much longer life than this. During this period, the system will have to be maintained; that is to say, corrections will have to be made to eliminate faults, modifications will need to be made to eliminate limitations in design and allow for increased volumes, changes in parameters, changes in business practice and so on. This is without the enhancements to the system which the user asks for to meet additional needs.

The effort expended in maintenance is far greater than that for the original development. Indeed, the cost of maintenance represents 80% of the total lifetime cost of a system – excluding running costs – over its complete life-cycle; this point is not normally fully understood or recognized. Any steps that can be taken to reduce this maintenance cost will thus be a material saving to the enterprise. A good-quality system that has been designed in a planned manner and according to a methodology that enables changes to be made will be more cost-effective in the longer term. Any additional cost during development will have a bigger payoff over the full life cycle of the system. A well-designed system will also, of course, reduce machine costs incurred in this maintenance activity.

Thus developing good-quality systems will not only provide job satisfaction for the development team but also ensure further development projects and save the organization money in the longer term.

10.3 Can quality be measured?

If quality is such an important component, then in the development of new systems, how is it measured? It could be easily argued that quality is subjective and cannot be measured at all, but adopting this argument means that the requirement for quality is never really seriously tackled and avoids the issue. We can, however, agree that it is difficult to define a quantitative measurement of quality, but unless some attempt is made to do so – no matter how rudimentary – quality will not improve.

A number of research papers have been written on the subject of the measurement of errors and much of their diagnosis is beyond the scope of this book. One measure of quality that we can use, however, has been defined as 'the number of faults identified per thousand lines of executable code delivered'. The number of faults is typically measured over a period of time such as in the first six months of live running; it can also be measured by a time interval, say by month, to give a fault identification rate. This rate can be plotted against time to give a graph which might yield valuable comparative information between systems. Indeed, this data is of very limited value in absolute terms but could be extremely valuable in comparative terms if measured for two systems developed by the same organization at the same time. Comparison would be thought-provoking and stimulate examination of why one project was of better quality than another.

Another measure which may be more relevant to commercially-based organizations is the cost of fault correction of a system against some measurement of the size of that system such as the volume of executable code. Similarly, the cost of fault correction per volume of code can be measured by month and plotted. This information can be compared between systems to establish why one is of better quality than another. This approach will also enable future maintenance costs to be estimated, an important factor in service organizations offering post-sales support and maintenance.

It has already been mentioned that a good-quality system has a longer life than a poor-quality one because the cost of ongoing support is less. The measurement described above directly measures quantitatively, errors present and subsequently identified in a delivered system; this is directly related to the quality of the system design. Quantitative techniques like these cannot be used, however, unless the systems design is supported by a formal methodology that determines in a consistent way the size and complexity of the system modules to be designed. Without some standardization in this way, comparisons become difficult to make.

10.4 Quality planning

A project development plan will certainly include a phase for systems testing even though it may vary considerably in its thoroughness. The tests and test data may be designed by the system builder merely to prove that the system does work as he intended, while a more thorough plan will be designed to establish the circumstances under which the system does not work as was intended. The testing process can then take place under the sole control of the system builder or under rigorous operational conditions.

More often than not, a system test plan is designed after the system investigation has been completed and while the programs are being written. Under these circumstances, system testing can only prove that the system does or does not meet basic minimum quality standards. It cannot, for example, ensure that high standards are maintained during the design phase even though this is highly critical to the long-term success of the system. Furthermore, any design weaknesses identified in system testing will prove expensive if not impractical to correct and this leads to the system being accepted as it is, whether the quality is good enough or not. In order to produce a high-quality product, it is essential from the outset that the system is designed with the conscious need for quality. This will involve the use of good design methodology and good-quality assurance procedures throughout the project life-cycle. To ensure that the system is designed for quality, preparation of a quality plan is needed.

The quality plan is a written statement of the methods used in the development process and the standards that must be met throughout the development cycle. The plan defines the quality required for the system not just in qualitative terms, but quantitatively too. For on-line systems, for example, the response time requirement will be defined. Without quantitative standards, quality can become no more than a good intention and not a practical reality.

A quality plan is thus an important step to ensuring good quality in the finished system. In the development of a system that requires high resilience, such as an embedded control system, then a quality plan is imperative. Other systems or projects where such a degree of resilience is not necessary or cost-effective still need quality plans, however, to define their quality standards. Even though the standard is lower, one must be aware of what it is. Without the plan, the danger is that the work on the second project is merely less thorough than the first and a slipshod approach ensues.

10.5 Quality assurance testing

10.5.1 The test plan

While the quality plan will define the performance standards for the delivered system, the quality assurance procedures check that the system, when built, meets the standards that were originally specified.

Quality assurance (QA) can then be defined as a planned set of actions and activities that will provide confidence that the product meets the established technical standards and gives the results and performance intended.

It is important to note that QA procedures are planned. Quality assurance is not just a set of checks undertaken when part of the system, such as a program, is ready; the QA activities must be planned in detail as part of the project plan.

The most commonly-used check is some form of software test. Quality assurance procedures therefore include a defined set of software tests with expected results, plus other steps such as reviews, inspections and audits, and are directly related to the development cycle of the project.

These appropriate tests are defined in the project plan and result from the functional decomposition of the system design. The project plan will have defined a statement of requirements for the project as a whole and for the sub-systems as appropriate; each sub-system is decomposed into further sub-systems or software elements that are themselves decomposed until the programs or program modules are defined. This decomposition of a system into component parts should be based on a formal structured design methodology. The number of sub-systems, the number of levels and the size of the final programs or modules will depend on the complexity of the project and the design methodology adopted. Some methodologies realize small cohesive modules that perform single functions; others realize more complex programs. This structure of the development project can be demonstrated by the diagram in Figure 10.1.

The tests that are undertaken as part of the quality assurance plan must not only check that each module does what it was designed to do but also the relationship between the module and any higher or lower module in the structure. The easiest approach then is to test out the modules separately and then deal with the tests that check the relationships between them. All modules must eventually be tested together to check that the whole system meets the requirements.

This process gives rise to three levels of testing:

(a) *Unit testing.* Where the program or modules are tested.

(b) *Integration testing.* Where the relationship between modules is checked.

(c) *Systems testing.* When the entire system or sub-system is checked.

If a diagram were drawn of the tests necessary to prove all the components and their relationships, then it would appear identical in shape to the structure of the development project. Thus, as we can see in Figure 10.2, the test plan is a mirror image of the design plan. Each design specification in the design plan should therefore be mirrored by a specification in the testing plan that defines the testing to be undertaken for that component. A test plan therefore has to be prepared for each component of the system whether this be a single module, a related set of modules, a sub-system or the system as a whole. This plan will show the number of tests to be undertaken, the objectives of each, the test data to be used and the results expected.

The logical point to begin the testing is with the lowest level of component in the system – the program module. The relationships between these modules can then be tested and, eventually, the whole system, thus working back up the tree structure. For obvious reasons, this is called bottom-up testing.

10.5.2 Module testing

Given that a good design methodology has been used in the development of the project, the lowest-level components will be small cohesive modules. Within each module, either a single function will be performed or the functions being performed will be similar in nature and be executed on the same element of data or similar types of data. This way, it is easier for an individual to comprehend the module

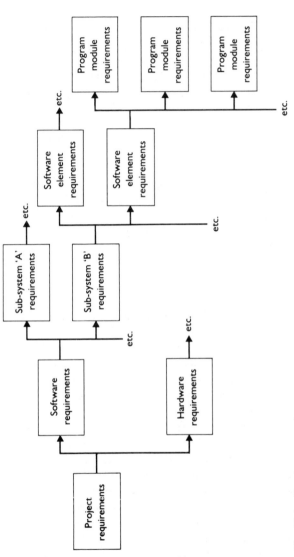

Fig. 10.1 Development project structure

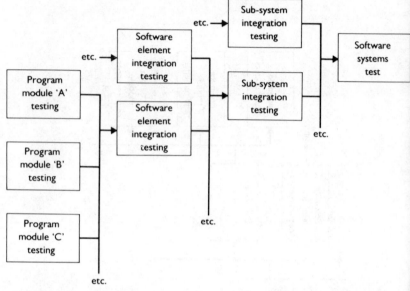

Fig. 10.2 Software testing plan

completely and write good code. It is feasible to design a test plan for the module that will traverse every logical path at least once and execute each statement at least once. The creation of test data will therefore be straightforward. The number of test shots will be limited and quantitative measurement of the errors found in each can be recorded. This is valuable feedback on the quality of the design and coding activities.

The expected results can be readily defined at this lowest level and the actual results checked against them. It is therefore relatively easy to identify practically all the errors in a module at this stage, and the expected results can be readily defined at this level and the actual results checked against them. If no defined design methodology has been used or if a methodology has been badly used, the module produced can end up as a program with several thousand statements for a number of different processing functions, handling many different record types. Under these circumstances, unit testing will be more complicated and will probably entail some form of simulated test data as it will prove to be very difficult for the tester to conceive all logical possibilities and very time-consuming to prepare all the test data. A test plan for such a program then becomes imprecise and is bound to include general statements such as 'test out such a function on the data until it works'. This is not satisfactory, and the subsequent task of defining the number of test shots and the expected results becomes quite impossible. The tester will probably have to test bits of the program at a time to make the task manageable, and the manner in which this is done is subjective and based solely on his experience. It would have been more preferable to have decomposed the task fully to give smaller, simpler modules.

Unit testing aims to eliminate all errors in syntax and to test the logic of the module. Time spent on thoroughly testing the lowest-level units according to a test plan will make the subsequent integration and systems testing much easier.

10.5.3 Integration testing

The purpose of integration testing is to check that the relationship between the lowest-level modules is correct; it is not to supplement the testing of these lower modules. Proceeding with integration testing without having completed thorough unit testing will undermine the subsequent testing and thus introduce unnecessary complications. It will be very frustrating when systems testing fails because of simple program-coding errors. There are a number of different ways that integration testing can be undertaken. Having written and tested all the modules that make up a sub-system, they can be tested together with, if appropriate, the control module. This is sometimes called 'phased testing', and would appear to be obvious and quite logical. It can have disadvantages, however. A fault may apparently lie in the interface between two modules or more typically between the control module and one of its subordinates, while it may have been caused by some interference from a third or fourth module. Thus the 'phased' approach to integration testing can make fault diagnosis more difficult.

An alternative approach, described as incremental testing, can be used. In this case, two modules are tested together to ensure that their relationship is correct; a third module is then added to check its relationship with the other two. If any error is found, it can more easily be traced as it can be assumed that the fault lies in the relationships with the third module. Once all three modules have been tested together, then a fourth module is added, and so on. If the addition of any module precipitates many errors, then it is possible to remove that module and examine it in detail to find out if it contains any fundamental illogicality. Having removed it, the original set of modules can be re-checked to ensure that nothing has been overlooked in testing so far. If the sub-system has been designed to utilize a higher-level control module, then an appropriate approach would be to test this, with its subordinates separately, then to test subordinates together with the control module and so on. When all modules have been added, then the complete sub-system has been tested; the same as if all the modules had been tested together in a phased manner. The advantage of the incremental approach is that it is far more likely to lead to a systematic and planned testing environment.

From the discussion that the test plan should be the mirror image of the development plan, the latter being a hierarchical tree structure, it would seem that integration testing should naturally work from the lower level of the tree to the higher. The lowest-level modules can then be tested in a phased or incremental manner to yield validated sub-systems, and finally these sub-systems can be tested together to deliver a proved total system.

There is an alternative approach to this 'bottom-up testing' – called 'top-down'. With this approach, the highest-level module is tested first and then the next level down, working down the tree. Top-down testing done in this way does not imply, however, that the lowest-level modules are tested last: these will have been tested as in unit testing. This top-down philosophy applies only to integration testing and requires that special test modules are written to simulate the subordinate modules of the one being tested. These subordinate test modules will be simpler than the real ones and will merely provide the correct communications with the higher module and will not themselves undertake any extensive processing nor access further subordinate ones. The only processing will be that required to simulate the required messages. These test modules are often called 'stubs'.

The protagonists of the 'top-down' approach claim the following advantages for it over 'bottom-up':

(1) The most important relationships are tested first. Not only that, they are tested repeatedly as the testing plan proceeds. In a real situation, all too often there are tight time-scales. With bottom-up testing, this means that the final stages of integration testing may be rushed, with disastrous results.

(2) A skeleton version of the system can be demonstrated early. This has considerable advantages in demonstrating progress to the user and showing what the final system will look like. This has, of course, its own dangers if the user finds the system is not what he thought he wanted.

but at least we know sooner rather than later. Making changes to the system during testing is very disruptive if no discipline over change is exercised. However, even more disruption and work will ensue if the system is delivered only to find there has been a fundamental error in design or a misunderstanding of requirements.

(3) A partial system can be implemented, if necessary. If, because of a particular problem, it is not possible to deliver the entire system within the time-scale, then certain subordinate components can be omitted and a simpler approach adopted – perhaps using clerical resources to achieve what the system should have done. This is not ideal but may be preferable to delaying the whole project until the problem is resolved. If the bottom-up approach is used, then a disproportionate amount of time may have been spent on the problem, thus delaying the whole project. Indeed, the problem may be caused by circumstances beyond our control such as a company re-organization which leaves part of the business in a state of uncertainty that will require some time to resolve.

(4) Programmer morale is improved. The development staff can see something tangible at an earlier stage and thus go forward with greater confidence. This also improves the relationships between analysts, designers and programmers, and they will work better as a team. With bottom-up testing there is often pressure on programmers to commence integration testing before they are ready.

There are, of course, some circumstances under which bottom-up testing may be desirable – such as when some of the lower-level components are critical in terms of machine performance. In these cases, the top-down strategy should be maintained and special arrangements made for these critical lower-level modules.

10.5.4 Systems testing

This is the next phase in the quality assurance procedure and follows integration testing once the relationship between the software components has been checked. Systems testing checks that the system does what is required of it. Its purpose, however, varies from organization to organization. It is a very general term used to describe the activities that follow program testing and often lacks clear objectives. In program testing, the individual programs will have been tested with their own test data to the satisfaction of their authors. What commonly happens next is that all the programs are linked together in a systems test. A number of runs are planned to simulate the live processing runs using data extracted from live files or created in some other way. This phase is thought to be useful as it will eliminate any remaining program faults because it is thought impossible to generate test data in program testing that would be comprehensive enough to exercise all logic paths. However, this approach leads to unnecessary complications and a protracted systems testing phase because the objectives of this phase have become confused. Many early errors in systems testing are caused by mistakes in setting up the systems test, errors in test data or errors in the interface between programs. It is only very late in this phase that we typically get round to seeing if the system is doing what it is meant to do.

We have previously seen that unit testing should eliminate as far as possible all syntax and logic errors in the module. Not doing unit testing properly and finding the same errors at the later stages of integration testing or systems testing will be much more expensive. A good design methodology will prevent this and thus facilitate this systems testing phase by ensuring that cohesive program modules have been designed. The integration testing phase will have tested out the interfaces between modules and their logical relationships; this testing phase will also have eliminated those set-up errors that bedevil the first runs of the systems test. This then leaves the systems testing phase to prove that the system meets the requirements specified for it.

The test plan is therefore prepared based on the separation of these objectives, and systems testing concentrates on proving that the system does what the requirements specification stated. This, of course, implies that the requirements specification will be used to produce the test plan and that the test will be planned to produce all the results at the appropriate point in the processing cycle. These results are defined in terms of screen formats or reports, the values of the output data and the criteria for acceptance. Expected values can, of course, only be defined if the test data has been planned in sufficient detail.

It may appear obvious whether the values are correct or not and, consequently, defining acceptance criteria may seem an unnecessary nicety. However, for all systems, let alone for the more complex interactive ones, it is essential to define acceptance criteria. Good project management will have ensured that this is done as part of the project plan.

A step-by-step approach is best for systems testing so that the most commonly used outputs are produced first. The periodically produced outputs – weekly or monthly analysis – will then be produced and then, finally, error condition or annual outputs. If integration testing has been thorough, the tester will not be worried about whether or not the system fits together but will be concentrating on its functionality. Similarly, if the first of the system tests concentrates on the basic functionality of the system, the tester will then move on to the more complex functions with some confidence that the system is in principle what the user required. Any faults in the interpretation of the more advanced functions will be easier to correct and not cast doubts on the whole requirement. This helps maintain good morale during what can be difficult times for the testing staff.

10.5.5 Systems test data

The same step-by-step approach should be used with the preparation of test data. The first runs should utilize relatively straightforward test data that will not invoke all of the functions and error routines. The testing for response times, throughput capacity and systems resilience should also be left to the later stages.

Ultimately, the final systems test will have to use some form of live or simulated test data. Typically, this is achieved by writing programs to extract data from live files to create test files and, at the same time that records and data structures are copied, various fields and codes can be charged to provide comprehensive test data. It is difficult to generalize about the preparation of test data as systems will vary enormously in complexity, size and functionality; and for larger systems the provision of comprehensive test data that will thoroughly test the system in a planned way will be a major task in itself. In these cases, the design, writing and testing of the programs needed to create the test data will have to be treated as a separate project and planned in the same way as for any systems project.

10.5.6 Operational testing

So far we have assumed that the unit, integration and systems testing have all been conducted in the same operating environment as the live system, using the same or compatible hardware and under the same software operating system as that to be used by the eventual user. It may be, however, that for practical purposes this is not the case and there will consequently be technical differences between the test and the live environments. Even in cases where the hardware and operating software are described as being compatible, it is worth checking this to avoid any last-minute pitfalls. There can, for example, be significant differences between versions of the same operating system that may affect processing.

Having recognized, then, that systems testing must be followed by a series of tests in the live environment, it is probable that the development team will have limited access to the live environment and it is, therefore, of paramount importance at this stage that we are not concerned with checking the functionality of the system. Any discrepancies or faults identified at this stage must be directly attributed to the change in environment to enable rapid identification and resolution. Similarly, in some cases there is some form of hardware that is an integral component of the total live system and this must be tested here with special attention to the direct interface between it and the software system.

The incorporation of this special hardware component in the total system should be the subject of a separate step in the QA plan and should be quite apart from any quality assurance procedures for the hardware item itself.

10.5.7 Acceptance testing

Having tested the system under test conditions to the satisfaction of the development team, it has to be handed over to the user. This may or may not have required operational testing, depending on the nature of the project. The user should then accept the system as meeting his requirements, accept responsibility for it and sign some form of document certifying that it is satisfactory. This may sound somewhat pedantic, but is a good principle to adopt and will avoid much argument and confusion later when any misinterpretation or misunderstanding subsequently occurs. In order for the users to assure themselves that the system is satisfactory, they need to undertake a series of tests, under their control and using their test data. The test data may be provided by the users' own staff inputting live or simulated data in volume, and in some cases the users may go out of their way to break the system by exploring all operational and logical possibilities; this is when a good systems test plan will have proved its worth. The purpose of this kind of testing is to prove that the system not only meets the users' requirements but is also resilient and easy to use.

The extent of this acceptance testing may well depend on the relationship between the development team and the user and the extent to which the user has been directly involved in the systems tests. Greater involvement usually means less acceptance testing. Equally, however, it may have been impossible for the development team to have simulated a live environment in their quality assurance test and so they may take an active role in acceptance testing. This is particularly true of large-scale on-line systems.

Therefore, it is not possible to be dogmatic about the relationship between the development team and the user during acceptance testing as it will vary from project to project and from organization to organization. However, in preparing the quality assurance test plan, due regard must be given to the need for acceptance tests as against systems tests and for the development team's role in such tests.

QA testing throughout all the stages described above will probably be necessary only on large or complex projects. However, the important thing is to recognize the particular testing needs of each project and have clearly stated objectives for each set of tests. Careful planning will avoid wasted effort here and pay dividends later.

With smaller commercial applications, the problem often lies in the establishment of requirements as the development of the system is technically straightforward, but even then a lack of planning in the QA procedures will lead to a poor-quality product being delivered. With larger-scale projects, particularly on-line systems, the amount of effort expended in quality assurance becomes very significant. For systems that require specified levels of resilience and fail-safe procedures, the testing effort becomes the greatest part of the total effort expended on the project. Under these circumstances it is essential that good-quality project control procedures are applied to the QA activities themselves.

10.6 Quality assurance reviews

10.6.1 Design reviews

In addition to software testing, there are other valuable activities that should be incorporated into QA procedures. These take the form of controlled appraisals by experienced staff of each element of work undertaken by a designer or programmer. The purpose here is to check that the work undertaken is in accordance with good practices and principles. These appraisals or reviews are meant to be constructive and engender good team spirit rather than to be critical of the responsible individual or group of people.

A design review is commonly a meeting or series of meetings of a group of people sufficiently qualified and experienced to comment upon a piece of design work. The architect of the work formally presents it and explains his rationale for the design; the audience will then offer criticism and alternative ideas for consideration. This is best handled by having a precise brief for the review, with the subject matter and its objectives clearly defined. This avoids generalized and abstract discussions on what is good design and what is not, and allows the group to concentrate on the subject being appraised and to avoid discussion on matters outside their brief. It is also preferable not to have management present as a design review should not be an appraisal of the individual's performance. Design reviews should be undertaken according to a schedule defined in the project development plan and should generally be held at each stage in the development process. The number, timing and nature of them will obviously depend on the nature and scale of the project and on the design methodology being used.

The following examples illustrate the process:

(1) *Systems analysis and requirements specification review.* The statement of requirements, or functional specification, prepared by the systems analyst in consultation with the user is in effect a design document. The review can check on its completeness and on the comprehensiveness of the specification to ensure that the requirements have been fully understood. The review will not, however, consider how the system should be designed to satisfy these requirements.

(2) *Systems design review.* This review will check the overall design concepts developed for the system, or for the sub-systems. This stage is critical to the project, and design reviews are most effective here if they deal with intellectually comprehensible pieces of design. Here, again, a good design methodology will help. An important objective of this review is not only to ensure good systems design in itself but also to ensure that it meets the detailed requirements or functional specification. The methods to be used in testing the design must also be reviewed.

(3) *Program design review.* This will review the technical specifications for the programs or program modules, or for a series of programs or set of modules. The purpose is to ensure good logical design and to check that it meets the higher-level systems design specifications. The program testing plans should also be reviewed.

From these examples, it can be seen that design reviews can be held whenever a design document or plan is produced. Reviews can be held not only for functional specifications, but also for project control plans and, of course, quality assurance plans.

The staff attending the review will not be the same for each one, but selected according to the level of the review and appropriate skills needed. There is no need for the project manager to attend each one and, indeed, he should not do so.

10.6.2 Walkthroughs

Another technique that has become more popular in recent years is that of the walkthrough. These are similar to the design reviews but may be called at any time in the project by an individual who would like to have his or her material reviewed. Walkthroughs involve the review of a piece of work by a suitably qualified group of reviewers under carefully controlled conditions. They are more appropriate to detailed level activities such as program design and code generation and are often used in conjunction with a structured development methodology where they are called structured walkthroughs. Other terms are peer group reviews or informal design reviews.

It is important to realize that these reviews depend for their effectiveness on a number of rules:

(1) The material to be reviewed must be complete and manageable: a program specification or a module of code are typical examples.

(2) The material must be distributed by the author some short time before the meeting.

(3) The meeting should comprise the author of the material, a reviewer who will act as chairman and someone to act as secretary to take notes; one or more additional reviewers should also attend. The review panel is selected by the author of the material from his peers or equals in the organization.

(4) The author of the material 'walks' the reviewers through the work step by step. The reviewers express any concern they have, the objective being to identify errors and check adherence to standards, not to offer alternative designs or criticize style.

(5) Any points that cannot be explained by the author are noted, as are all the action points.

(6) A further meeting may be called by the chairman to follow up the action points.

(7) The meetings should be brief – no more than an hour – so that they concentrate on error identification.

(8) It is the responsibility of the author to take the appropriate action and report back to the panel.

(9) The results of walkthroughs should be filed for future reference.

The emphasis of walkthroughs is that the reviewers should be supportive and the author should not feel defensive about any errors identified in the work. For this reason, it is inadvisable for managers to attend!

10.7 How a structured development methodology improves quality

Throughout this chapter there have been many references to the value of using a good design methodology with respect to quality. This is not the place for a detailed discussion on the benefits of using a structured system development methodology nor for a comparison of the various methodologies available. However, it will be useful to summarize the situation and highlight a number of points that are particularly relevant to ensuring a good-quality system.

Structured development methodologies all attempt to solve the problems of systems development and project control by the application of a disciplined and organized approach based on the principles used for product development in engineering industries. Hence this gives rise to the expression of 'software engineering', although this term implies the use of other techniques apart from structured methodologies.

Structured methodologies all attempt to break down or to partition the system under investigation into a series of related logical activities or actions. The scope of the system and the interface to the outside world are clearly defined. The logical relationship of the activities or actions is described by the use of diagrams and charts in a nested or hierarchical structure. The actual tools used, that is to say the diagrams and charts, vary from one methodology to another but are all basically similar. A structured methodology will also define the steps to be undertaken in the systems development life-cycle, the sequence of these steps and the tools to be used at each stage. The methodologies encourage the use of iteration in the development of a design and can provide alternative design options for consideration by the user. The system is investigated and designed, based on logical considerations, ignoring the limitations of computer hardware and software. This makes it far easier to involve the user in all stages of the development, and a key benefit of these methodologies is that users are involved on a planned regular basis throughout the development life-cycle.

The details of the logical processing are described using diagrams and charts, tables, pseudo code, structured English etc., depending on the methodology. The need for imprecise narrative is minimized and the creative thinking process is carried out in a formalized manner, encouraging due consideration of all the factors and aspects involved. When the system has been analysed and the solution to the problem defined in logical terms then, and only then, are the physical characteristics of the hardware and software taken into account. This avoids making the problem fit the solution – the weakness of technology-driven solutions. There are, however, some limitations and drawbacks to using a structured methodology: some claim it inhibits creative thinking, others that it is not always practical in a commercial environment.

There are two basic types of methodology – data-driven or function-driven, both with their adherents. There are also two approaches to the use of the various techniques employed in a methodology. The first, often associated with a function-driven methodology, can be described as the 'toolbox' approach. Here, the emphasis is on recommending the use of alternative techniques for certain stages of the development cycle. The developers select the tools with which they feel most confident and which are most appropriate for their development environment.

The other approach can be described as that of the 'cookbook'. In this case, the steps to be undertaken and the techniques to be used are precisely defined for the stages in the development cycle. The composition or ingredients of each stage are precise and it is a requirement to complete all activities before progressing further. This approach gives less latitude to the developer but is more disciplined. As one of the objectives of using a methodology is to introduce more discipline, the cookbook approach is often preferred, and is undoubtedly better for large-scale projects; for smaller projects, a simplified version can be used.

Many systems delivered by traditional methods are often, but not always, of poor quality. Typically:

- The system does not meet the user's needs
- It is too complicated to use easily
- It is not reliable
- It is too difficult to modify
- It costs too much to maintain

Apart from the many problems encountered in project control of the late delivery and excessive costs of development, the use of a structured methodology should ensure that:

(1) The system boundaries are clearly defined so that the scope of the project is understood by both user and developer.

(2) The system is logically partitioned in a top-down manner, reducing interface problems for subsequent enhancements.

(3) The steps to be undertaken in development are defined, thus facilitating project control and QA procedures.

(4) The user is more easily involved in each stage of development, and alternative designs can be modelled.

(5) A modular system is evolved based on logical data or functional relationships, each module having a precise function and data integrity.

(6) The solution is based on possible data structures and data flows and is not initially constrained by hardware or software limitations.

(7) It is possible, if a top-down design approach has been adopted, to implement partially developed systems.

All the above, and other benefits, will give rise to a better quality system because of the more organized and disciplined approach adopted by the developer. There is also a more direct effect on quality: the defined stages of development will enable milestones to be specified at which QA checks can be undertaken. As these will be the same for all projects, the QA procedures will be the same and will not be overlooked. The exact outputs from each stage can be specified by the methodology and these, in the form of charts and tables, can be inputs to QA procedures.

The key to successful design reviews, structured walkthroughs, code inspections and the like is that the matter under examination should be readily understood without excessive effort by those not directly involved with that particular task. The parcels of work must be complete, not too large or complex and have clear boundaries; any open-endedness will invalidate the walkthrough or review.

The true modularity provided by structured systems development methodology will greatly enhance the effectiveness of walkthroughs; indeed, some will argue that a walkthrough is valueless if a structured methodology has not been used.

The partitioning of the system into sub-systems with clear boundaries and the modularity of the system will make the planning and preparation of systems test data much easier and much more effective. Similarly, unit testing or program module testing is much more effective if the modules are logical entities performing one function or processing one data element.

As the user has been involved in the design, the final system will not be a surprise to him, and the user acceptance testing phase will concentrate on proving the system. The user's efforts will be devoted to the original purpose of the testing and not on arguments about the misinterpretation of requirements etc.

10.8 Summary

At the beginning of this chapter, we discussed what is meant by quality in the context of computer systems by drawing an analogy with the quality of any durable product, and concluded that a good-quality computer system is:

Error-free
Complete
Comprehensive
Reliable
Easy to maintain and enhance
Flexible
Efficient

We established that developing good-quality systems ensures that the development team continue to undertake further development projects and that they provide financial savings in the long term.

We discussed that although difficult, some form of quantitative measure of quality was important and that this was of most value in comparing the quality of different systems.

The next section explained that to ensure quality, the development methods and standards required must be defined in a quality plan for each stage of the project.

This was followed by an explanation of QA procedures, which were defined as a planned set of actions and activities that will provide confidence that the product meets the established technical standards and gives the results and performance intended, the most commonly used form of checking being the software test. We recommended that there should be a clear separation of the objectives of the software tests into:

(a) *Unit testing.* Where the program or module code is checked to prove the logic and syntax.

(b) *Integration testing.* Where the interface between modules is proved.

(c) *Systems testing.* Where the functionality of the system is proved.

This may be followed by operational testing, depending on the nature of the project, and by acceptance testing. There would inevitably be some overlap between these two, depending on the organization; the important point being defining the objectives of each of these tests so that fundamental errors did not disrupt the subsequent ones.

The next section explained how design reviews and walkthroughs improve quality and how they should be conducted. The key points being that they should be undertaken in a disciplined manner, the objective being the elimination of errors rather than the criticism of individual authors.

Finally, we reviewed the importance of using a good structured development methodology in ensuring good-quality systems. In particular, how a methodology will create logical modules which will greatly facilitate QA procedures. The key message must be that the quality is important and that it must be planned for; it does not happen by goodwill alone.

11 Managing the implementation

11.1 Introduction

Implementing information technology-based systems is never an easy task, no matter how much technical preparation has been carried out nor how technically skilled the project team. The complementary factor here is good management. It is worth remembering that the traditional career path for information systems staff is based on the demonstration of technical knowledge and practical excellence. Suddenly, however, the systems analyst is thrust into the role of a project leader responsible for an implementation programme and all of the management problems that brings. New perspectives then have to be adopted and additional skills learned. For these reasons, this chapter highlights the need for an understanding of the process of change both within the project team and in the organization as a whole. In particular, it deals with the way new technology can have an impact on people, jobs and the organization.

11.1.1 Managing change

The management of new systems implementation cannot be treated in isolation since it is a consequence of other, perhaps more fundamental, changes within the organization, such as a new way of conducting the business or a different philosophy about productivity in the enterprise. Such changes of philosophy or direction are almost invariably bound up with new technology and will have to be managed, together with the technical aspects of the system. It follows, therefore that understanding the process of change is a new critical factor in the successful implementation of new systems.

Change does not occur in a vacuum. An enterprise has a long memory and the past history of systems implementations will condition people's attitudes and commitment to the new technology and the project team. Failures or less than perfect implementations are especially remembered. The kinds of comments which may start to surface are: 'This is the fourth time we've tried to use new technology in this area! Why should it work this time?' *or* 'The last time we tried this, I spent five months preparing my department and it all came to nothing. I can't ask my people to go through all that again!'

The project leader must take time and trouble to collect these views, however irritating and unjustified they may seem. That way, a strategy can be developed to counter them and make explicit the positive and negative aspects of the implementation so as not to generate false expectations. Hiding the possible deleterious effects of systems change serves no purpose – hence the need to be open about the effects of new systems which might be considered negative. A balanced discussion can often lead to a more effective solution and to a more creative design which minimizes adverse social effects.

Apart from the general culture and belief about new technology in an organization, there will be all kinds of vested interests surrounding the system. Even if everyone endorses wholeheartedly the broad benefit to the organization, such as 'it will give us a competitive edge', people are unlikely to agree completely on the means by which this should come about. After all, the staff all have different jobs to do even though they are working towards the same organizational goals, and it is unlikely that all jobs will be affected positively by systems change.

As well as considering the context of change and the differing perspectives from which the system will be viewed, the project leader must acknowledge that the acceptance of change within different parts of the organization may not be as rapid as enthusiastic supporters of the system may want. The pace of technologial change and attitudinal change are unlikely to correspond: technical awareness programmes and training may impart knowledge and skills; attitudes take longer to change and can require a totally different approach, phased over a longer time period.

Finally, two notes of caution. First, poor management practice will be exposed either by the requirements of the new system or will be reinforced by it as users attempt to shape it to their own particular needs. Such divergent behaviour will cause implementation problems. The project leader needs to have a clear view of the objectives of his new system and of those it will serve if he is to overcome this. Secondly, the process of change does not always proceed in a linear direction or logical manner: people may appear to take a step forward in their understanding or commitment to the system and then change their minds at the last minute for no apparent reason. The project leader needs to spot when this is happening and take appropriate action.

11.1.2 Balancing needs

The implementation process requires a careful balancing of the needs of the enterprise, the people and the technology (Figure 11.1). Maintaining this balance must be a prime concern of the implementation team.

It should be remembered that the client or group who sponsored the new develoment will have been convinced to back it only because of a conviction that it would bring major benefits to the enterprise. Every opportunity should therefore be taken during the implementation to reinforce this belief so as to maintain momentum and morale. All employees, not only the board and top management, are able to identify and recognize the need for improved customer service, responsiveness to changes in the market place, and so on.

However, over-emphasis on the needs of the business and technology at the expense of the people in the organization will lead to a sense of dissatisfaction and alienation among the users. Expressions such as feeling like 'a cog in a wheel' or 'the inhuman face of technology' best sum up the end result. On the other hand, if the system designers and implementers place too much effort on satisfying user and technology 'friendliness' criteria, this may not meet the business requirements. Finally, if business and user needs are placed before technological requirements, change may be seen to require too much effort; this is

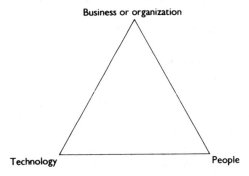

Fig. 11.1 Balancing the needs of the business or organization, technology and people

often the case where the enterprise is successful and sees no need for change. Alternatively, the organization may become totally unrealistic or uncompromising in its expectations and expect technology to provide everything or nothing.

11.2 The implementation process

In an organizational context, the implementation process can be split into six stages. While other structures are used, the list below focuses on the process as experienced by implementers and users of technology, and treats implementation as a complete change process divided into:

(a) planning the change;
(b) considering the implications;
(c) developing the system;
(d) evaluating success;
(e) expanding into new areas;
(f) consolidating change.

(a) *Planning the change.* At this stage, the organization has probably been waiting a long time for the promised new system. Pressure may be building up to 'go live', and the temptation for all concerned may be to rush their fences. However, this is exactly the point at which the project team must take stock of some of the critical issues.

● Have the business needs changed?
● Are the objectives of the system still the same and expressed in the same way?
● Are there any new constraints to be taken into account concerniing the use of equipment or software?
● Have the lessons from the prototypes or field trials been properly understood?
● Do the time and cost projections need revision?
● Are any job losses anticipated?

All these and many other questions will need to be answered to ensure that there is a definite baseline from which the change process can start, and this will ultimately be crucial in evaluating its success.

(b) *Considering the implications.* When this stage arrives, it is important to feel confident that the organization as a whole and the function being supported by the new system have been prepared for the coming changes. While it is not possible to predict events exactly, some general systems education and orientation training should have been carried out. Without causing a state of alarm, people should be encouraged to look for possible effects that the new system will have on their jobs. While it is unlikely that most organizations will be able to cope with such an open view at all levels, management should be expected to treat the process of systems change in a realistic way and be prepared to weigh its consequences. A good liaison procedure will need to be established so that adverse comments and suggestions can be picked up early by the team. As far as the projct team themselves are concerned, all their roles and activities should now be clearly defined so that they have some structure within which to work and some milestones to work towards. They should also have made explicit their own model of the end user for whom the system will be prepared. The skills or aptitudes the users possess and their general attitude towards the proposed system will need to be considered during the change process.

(c) *Developing the system.* It is unlikely that the systems project team will have much choice in deciding which of an enterprise's systems should be tackled first. As we have seen in Chapter 1, the development of business plans and their derived systems plans will have determined the likely starting point. Influencing this planning process, however, will be the systems team's wish to start where they can have the greatest chance of success. This might be:

- where users are most enthusiastic and where high motivation can be used to fuel the process of change;
- where the project development time will be short and where a success can be demonstrated early: there is an old saying that 'nothing succeeds like success' and a good early implementation boosts morale all round;
- where the effect of the change will have high visibility and make a significant contribution towards helping the enterprise to achieve its goals.

(d) *Evaluating success.* An early implementation needs to be closely followed by a short period of evaluation. Questions which the team can ask are:

- Did anything unexpected happen?
- Did we react in the best possible way?
- Have we placed an unfair burden on the end user?
- Did we give the end user enough opportunity to influence what was being done?

● Is the users' commitment increasing? Are they constantly volunteering suggestions?

The answers to these questions need careful analysis before expanding into new areas.

(e) *Expanding into new areas.* By this time, a momentum will have been built up and, if the previous implementation has been successfully completed, there will be pressure to keep up the momentum. Timetables and procedures will need to be re-checked and revised in the light of experience, and a planned campaign of internal publicity will need to begin. Here, the commitment from top management will need to assume a high profile in order to keep resources available.

Against this back-drop, further implementations will proceed. During those times when morale may be lowered due to some temporary crisis or setback, it is this baseline of commitment which will keep the project moving. When people cannot see success immediately or when implementation priorities clash with normal work, it will be the commitment of top management that will keep the project as a high organizational priority and ensure that necessary resources are available. The impact of the new systems on the enterprise and its way of working will now begin to be felt.

This planned expansion of the new system may continue for an extended period, and a great deal of psychological and emotional energy needs to be released and channelled to keep the stream of new projects alive. Any wavering of commitment or casual remarks of concern from key people will quickly be interpreted in other parts of the organization as signals of dissatisfaction. Thus it is crucial to acknowledge a crisis if it exists but at the same time to communicate very positively how it is going to be tackled. If successfully handled, the team can capitalize on this to reinforce confidence.

By the same token, it is important for the project leader to know how to maintain morale within the implementation team itself. They need to realize that they will become the focus of possible uncertainties and worries of the end user. The end user manager will need to play 'devil's advocate' to an almost extreme degree in order to have the answers which he can pass on to his own staff who may need convincing and reassuring. These difficulties will be increased if the end user manager is struggling to clarify things for himself – so it is worthwhile ensuring that he understands everything from the very start.

(f) *Consolidating change.* After a while, new systems begin to bed down and become easier to control internally. It is often the case that the tracing of errors by the end user can be a means of deepening their understanding of their organization and the relation of the new system to it. This can occur, for example, with systems where information may need to be obtained from several different departments before the appropriate end user information can be collated into a usable form. For example, the preparation of invoices for goods that are to be exported may require the integration of information from sub-systems, such as details of items selected from a warehouse, freight charge ratings, credit authorization and so on.

This is also a time when valuable information can be gained from the users regarding improvements and possible new developments as the wider implications of the new system become more obvious. Changes may now need to be made to the old organization structure and reporting arrangements as the new system may cut across old boundaries. If not recognized at an earlier stage of system design, this situation may come to light in clashes between job-holders regarding their responsibilities and accountabilities. For example, personnel handling customer enquiries may find they are able to deal more effectively with them because they have more relevant information provided by the system in a speedier manner. Because they appear to be able to satisfy the customer demand for information, they may feel under pressure to make decisions. It may then be appropriate for the organization to re-think its policies and practices. For example, extra responsibility may be given to staff allowing them to authorize refunds on goods within pre-defined limits or to give out information on delivery time for goods; previously, this information and authority may only have been available to supervisors or a similar level. If the new technology is also for direct use by customers – e.g. a viewdata service for travel agents to book holidays – then a special unit may be necessary to monitor and support their activities in relation to the business. As the benefits associated with the system become apparent, they should be widely publicized as this will confirm the wisdom of embarking on this course and prepare the way for future development.

11.2.1 Fundamentals of management

The implementation of change is a testing and complex process which needs to be managed. Definitions of 'management' abound, but two which are very relevant in this context are 'getting work done through other people' and 'planning, motivating, co-ordinating and controlling the activities of others'.

The first definition is of particular interest here since few systems assignments are so small that they can be done by one person. A mixture of skills, abilities and personalities is usually needed to ensure that all the aspects of a project can be covered by the team. At the same time, the people in the project team have needs of their own: achievement, personal development, a favourable self-image, acceptable co-workers, and a sense of belonging and identity with the group. A manager who uses people – in the very worst sense of the word – and then discards them or ignores their needs after they have completed their part of the project is not keeping his side of the 'psychological' bargain. During the implementation, especially at crisis points, meeting these needs may become a critical factor to the success or otherwise of the project.

The second definition is really an expansion of the first, but highlights four activities crucial to successful working. First, planning is obviously essential for implementation, involving as it does a whole host of different elements: the right infrastructure and environment for the technology (such as space, cabling, modems and so on); the correct phasing-in of resources; and the purchasing of the most suitable hardware and software. Small omissions of detail can have disastrous and disproportionate 'knock-on' effects throughout the whole project, causing delays to the project team and users alike. However, planning is not a 'one-off' activity carried out at the beginning of implementation: it has to be done continually as there will always be an opportunity or a crisis which will cause revisions to be made to the plan, as we have seen in chapter eight.

Motivation is the process by which the project leader makes people feel that their own needs coincide with those of the project. This can be seen when team members show initiative, take on extra responsibilities, are willing to work overtime or identify difficulties without taking the attitude that 'it is someone else's responsibility' or 'if I mention it I'll have to put it right'. Motivating people can be a difficult task: where, for example, there has been a long lead time from planning to implementation, the team, if its composition has not actually changed, may have 'gone off the boil' and the necessary mental energy may have been dissipated. On a very large project, people may become extremely bored if they have to repeat a particular phase several times at different sites. While they do this and build up knowledge and experience valuable to the team, they may be alienated unless they are moved to a different task which satisfies their need for challenge.

Lack of some immediate results can be as frustrating from the project team as for the end user. It can lessen enthusiasm and an early failure can be devastating to morale. Even though the cause of the failure may lie in the hardware or software, in the user's eyes it is still the project team that has failed. The project leader will then need to restore confidence not only in the team but with the users and the rest of the organization. If an implementation is taking longer than expected and no benefits are shown – or even worse, the new systems perform less well than expected – the project team may find itself the target for implied or direct criticism. We shall return to this later in the chapter.

The third aspect of management, co-ordination, naturally becomes more difficult as the size of the project increases or the applications involved become more complex or diverse. Co-ordination means using at the appropriate time all the right human, physical and technical resources; these may all have different lead times, of which the project leader must be aware. Attitude change and skills development, for example, cannot suddenly be effected, unlike a technical component or package which may be rushed through a delivery cycle. Once brought into use, the interactions between these resources need to be monitored.

Fourthly, control procedures will have been defined in advance of implementation. Ensuring that these are adhered to, however, can be very difficult, especially if a project falls behind schedule and users are impatient for results. Moreover, while a project leader may control people's time, their aspirations or other opportunities for career development which come along are not subject to this control. Therefore, changes in personnel, especially key staff, and the possible effects this could have on the team and its progress need to be anticipated. In addition, certain personal characteristics may dominate an individual's method of working, and the project leader has to be alert for the 'technical enthusiast' as much as for the 'cutter of corners' as behavourial extremes will damage the team's chances of success.

Controlling the timing and pace of the project is always difficult as factors will arise beyond the project leader's control. As we saw in Chapters 3 and 4, breaking up the project into manageable milestones will at least give the team a target to aim for and enable them to better assess the time needed for tasks if the schedule has to be re-calculated. The nature of the control process is, however, a subtle one.

One interesting aspect of control is that it is always something other people need! However, the mere fact that a controller or manager is there to ask questions will ensure that another perspective is brought to bear on the systems implementation. Furthermore, there is a difficult balance which needs to be struck between over- and under-control, and we can now examine some of the causes and consequences of these two extremes.

A manager may over-control a team because of insecurity about the eventual outcome of the project or because the consequences of failure are severe at a personal and organizational level. The response of staff may vary in form but can be detrimental to the success of the project, bringing about the worst fears of the project manager. For example, over-control discourages the healthy questioning of what is being done, suggestions for improvement and the taking of initiative. Staff may also interpret it is a lack of confidence and trust in their ability and integrity as professionals. The team may then not feel responsible for what happens, always 'passing the buck' upwards. In extreme cases, they may retaliate by referring every small decision, bombarding their bosses with memos and queries in order to overload their manager, thus diverting attention from the main task; they may also withhold relevant information, ultimately ensuring that there is no adequate control. Under-control can be equally destructive: in this case, the manager may be too disorganized, uninterested or busy; he may dislike bureaucracy and giving the necessary guidance. The staff may then become disorientated through lack of clear directives and deadlines. If the yardsticks are not applied, there is no feedback, which is itself de-motivating. The team may also underestimate the time and effort needed to do their tasks and be unaware of possible pitfalls. Finally, individual team members may focus on things that interest them at the expense of the main job – the implementation.

11.2.2 Who will manage the whole implementation?

It is nearly always assumed that the information services department will provide not only the project leader but the manager for the whole project. However, the time is fast approaching when this may not be a foregone conclusion. First, systems personnel may move into line management positions, especially if their internal promotion has been blocked or they can see better opportunities in related commercial enterprises (e.g. managing the marketing or production of systems-related products or services). They may therefore be very suitable candidates for managing the project.

Secondly, there is a growing body of more sophisticated end users which has used information technology and has a good technical appreciation of what a system may be expected to do. In some cases, it may be a valid question to ask whether the manager of an end user department should manage the entire project. The advantage of an end user manager in this role is that the primary concern will be the contribution of the system to divisional or departmental effectiveness since this is how personal performance is measured. Therefore, there is an over-riding commitment to make the system work, which will be communicated to the staff; all decisions taken will therefore be subordinated to this one goal. In this situation, the information services department is a specialist, expert resource and does not have ultimate responsibility for obtaining user commitment – it is the user manager's job! It also means that if the end user manager understands the system, the staff will receive more information direction and support. A disadvantage with these arrangements may be a 'conservatism' as to what may be achieved, and new system developments may be limited to what can be achieved in the short term. The systems personnel may consequently feel that their power is being eroded or their skills transferred. However, if their contribution is judged by the successful use of the system, then an end user project manager will be an enormous asset in achieving this.

If the project manager comes from the information services department, the advantage is that he or she knows what is technically possible and can push the performance of the application to its limits, conversely, there will not be unrealistic demands about what the system can or should do. But, being removed from the day-to-day practicalities of the user department's business, systems considerations and convenience may unconsciously come to dominate decisions because the user is not technically sophisticated enough to argue the toss or to be entirely sure what his department is being expected to achieve. Last, but not least, there is a strong possibility that the project manager will become too immersed in detail.

11.2.3 The criteria for success

The ultimate criterion for success will be whether the system is used successfully and is therefore contributing to overall business performance. For this reason, the attention that must be paid to the end user cannot be overstressed. Of course, there will be many other criteria for evaluation, depending where in the organization an individual sits. The accountant, for example, will be concerned with time-scales, costs and payback of the investment; others may emphasize speed of transactions, reliability, good maintenance and so on. In the end, all these aspects are negated if the end user ignores the system. That is why their training and involvement are so crucial, as we shall see later in the chapter.

11.3 Why implementations fail

The implementation of new technology may fail for a number of reasons, many of them non-technical! The potential benefits in relation to the business or organization may have been wrongly assessed; there may have been an inadequate analysis of the costs and risks associated with the system; there may have been little support from top management and a corresponding lack of resources – human, physical and financial; the feasibility or prototype studies may have been too superficial.

A key reason for failure, however, often relates to the lack of user acceptance; and this can be seen at all levels in the organizational hierarchy. Formerly, it used to be thought of as essentially a clerical or operator problem with large batch data-processing systems. Now, new interactive systems built around individual workstations offer many different facilities to all kinds of professional and managerial staff at all levels of the organization up to main board level.

Lack of user acceptance can also be seen horizontally, right across the organization – marketing, sales, production, quality control and so on. Sometimes, more subtle effects are only noticed by examining the relationship of one department or division with another. A system to improve a factory's productivity, for example, can cause problems for the indirect user, such as the marketing department, which now faces the problem of disposing of goods without lowering prices and starting a price war. The success of a sales force boosted by technological aids can become the distribution manager's nightmare! User acceptance must therefore be seen in a broader, organizational, context, with a view to balancing individual and departmental needs in relation to overall business performance.

A third consideration is the type of job the technology supports. Secretaries producing vast quantities of text, mechanical engineers designing complex equipment and financial planners may all, in their turn, be sceptical of word processing, computer-assisted design and financial-modelling packages which are offered to them and, perhaps on occasions, with some justification!

All these different occurrences of lack of user acceptance may smoulder away while system teams are working enthusiastically at the technical aspects of the new system during implementation. The problem may therefore only receive recognition once a crisis point has been reached: for example, when a million-pound network of multi-function workstations lay idle because people refused outright to use it, saying it did not improve the way they did their work. Curiously enough, all the technical training had been carried out (that is, how to operate the system) but its application to business problems had been neglected and managers were too busy to stop and make that connection themselves.

11.3.1 The cost of failure

Failures can be very costly and their full effects may take some time to emerge. Two main damaging results are:

(a) the effects on the business (such as dissatisfied customers, lost business, missed opportunities and extra pressure from competitors);

(b) internal effects (such as wasted resources, de-motivated staff and loss of key personnel).

The effects on the business may only be recognized some time after the event, when it is too late to do anything to rectify them. Moreover, they may trigger off secondary costs, such as extra effort to placate customers or additional advertising expenditure to compensate for missed opportunities. Poor business performance, in turn, will sharpen the effects on the internal organization, and the whole process repeats itself.

One way of avoiding these hazards or, at least, controlling or lessening their impact is to focus on the characteristics, needs and interests of those who are likely to be affected by the new technology, paying particular attention to the impact that the system may have on the jobs people are asked to do, and on the organizational climate and structure in which they operate. Training plays an extremely important role in this and we turn our attention to it now.

11.4 Training

It is at the implementation stage that the soundness of the training strategy will be manifest. In order for training to be successful, certain issues will need to have been clarified:

- Who needs to be trained and why?
- Will all the training costs be borne centrally?
- Which departments, if any, have allocated part of their budget to cover training costs?
- Will training take place on or off the job?
- Who will be involved in training?
- Should this training be a long- or short-term investment?
- How will line management be involved?
- Have the hidden training costs been assessed?

If these questions have not been answered, thrusting a system manual, however comprehensive, into the user's hands or picking the most 'user-friendly' menu-driven technology will not guarantee success! The reason is that training is not something done to the user; it represents the culmination of the decisions of many other – sometimes competing – interests within the organization and must be carefully assessed while plans for the new system are being prepared.

First, enlisting the support of the end user manager is vital since he or she will decide how many people can be trained and at what time, without lowering the department's performance to an unacceptable degree. In some cases, standards will have to be relaxed while users become proficient in the new system. Another role for that manager is acting as a gatekeeper and diplomat between the department or group and the rest of the organization. If, for example, the department has not provided the expected level of service or if errors arise due to the new system, the patience and co-operation of others may be needed in tracing them through a function or process so as to identify the fault.

Secondly, training needs to be carried out in a way which will achieve results most effectively. One approach is to send a few key people in the user department on a course and then have them teach the rest. The advantage is that these people understand the nature of their own work best and can translate their newly acquired language and skills into the concepts and language familiar to the rest of their departmental colleagues. They will be more sympathetic to the real, or imagined, difficulties that their colleagues have and the latter will, in their turn, have confidence in their instruction. Finally, the whole process of disseminating the training is gradual (trainees can volunteer) and therefore less of a culture shock to the department. There comes a point when a 'critical mass' have been trained, and those who have not become anxious lest they be left out. The costs with this method are therefore minimized as most training takes place in the user department.

Alternatively, everyone may be sent in large batches, closely following each other, on a course with vendors or training specialists. This can be time-consuming, costly and very disruptive to the work of a department: a short, sharp shock! However, when people return to their job, there should be enthusiasm for the new system and everyone is expected to know what they are doing.

A third possibility is that an in-house resource will be made available for training. If the systems department and training department are too heavily involved in end-user systems training, other commitments may suffer; there is also the danger that they may be seen as too remote from day-to-day operations. Finally, if their knowledge has been acquired from vendors or their specialists, they might place less emphasis on or even omit areas critical to the end user's success with the system.

No single solution is likely to be perfect nor will the same solution work in each case. Also, much end-user training can now be delivered to the user via the same hardware used in the system through the medium of computer-based training (CBT) courses. This trend can be expected to increase as CBT methods become easier to use and as artificial intelligence approaches become more widespread.

11.4.1 Content

Just what should be taught is a matter for a careful training needs analysis and a rigorous specification of instructional objectives. However, this is a subject in its own right with a large literature, and the advice of a training specialist will be invaluable to the systems team. There are three broad training areas which should be covered to achieve a successful implementation:

- orientation training
- application area training
- 'hands-on' training

Orientation training should already have been carried out before implementation has begun. If it has not, then it is still not too late, but the benefits of doing it earlier have been lost. This type of training should aim to create the right cultural environment for the system and provide a context in which the application can establish a firm foundation and grow. It should consist of an appreciation of the effects technology may have on the business; the different types of technology and the way they are inter-related; how information is stored and processed; the physical security of data; what happens to data when the system is down; how software can protect the system from user errors by inbuilt checks. Then there are basic housekeeping rules and the protection of individual privacy. This type of training will enable staff to contribute more fully to the implementation process.

Application area training is the natural successor to orientation training. Once a general understanding of information technology has been gained, the organization will be ready to absorb more detail about what specific application is designed to achieve for the business as a whole and the particular organizational unit selected. It should cover the relation of the systems to other systems and to other areas of the business and their associated divisions or departments. This is vital, since any system will affect and be affected by activities in other parts of the organization. Often, an end user may become aware of how the system links to other parts of the business only when it is 'down'. This type of training will lead to a better understanding of how individual jobs may be affected and how different departments must fit together to ensure a smooth operation.

'Hands-on' training refers to the actual operation of equipment by the end user. There are many ways to involve the user: by having simulations and games or by inviting the user to try to beat the system. This increases the understanding of the system and the confidence in both the user's ability and in the resilience of the system. For example, some users may be fearful of experimenting with the system in case they make a mistake which could result in great expense to correct. This type of training is even more crucial where the means of data entry and control is not via keyboards but uses the mouse or the light pen separately or in combinations, as in some computer-assisted design workstations. Computer-based simulations and computer based training courses have an important role to play here.

11.4.2 Methods and media

Once the basic content of training has been decided, the question is raised of how best to put the message across. The least useful – but still used – method is to give the user an enormous manual which he is expected to read through systematically and to emerge from with infinite confidence. The user is, of course, entirely rational and logical, will retain all the detail and not become bored!

The reality is that human beings do not learn in this way. Learning is more effective if material is broken down into manageable 'chunks', with an opportunity to apply it regularly in relation to the system, and at suitable intervals. Suitable in this context means a period of time long enough for the learner to consolidate information before he or she forgets the knowledge acquired. Practice with the system will itself generate feedback in the form of errors, questions and uncertainty for the user, which will then influence how the subsequent parts of the manual are approached and digested. What happens most often is that the learner is expected to understand the manual and carry on working. Fluency in the use of the system never builds up because of the constant need to stop and refer back to the manual. Learning must be supervised and controlled; if it isn't, the learner may appear to operate the system competently but, because understanding is incomplete, additional practices and procedures may start to grow which eventually become counter-productive.

Other teaching methods such as individual tutoring, on or off the job, or supervised working of groups allow for a more personalized instruction programme and for more immediate feedback. However, this can be costly and adminstratively more difficult for the organization in the short term. The project leader should ask what the preferences are for each user department.

Potential media for training have now become more exciting both for users and trainers alike, with the advent of VCRs, computer-based training and interactive videodisc. The last offers many new possibilities for the user in that direct access to material, remedial exercises, quizzes and audio-visual presentations can all be provided. Moreover, several types of course can be made available at different grades of difficulty and in different languages to cater for variations in sophistication and culture of the user. The advantage of interactive learning is that students can pace themselves and use the computer-learning system at times convenient for themselves and their job; instructors' time is saved, especially if there are many people to be trained in different locations. Another advantage is that records can be kept of the types of errors made by the learner, the number of times remedial frames are accessed and so on. All this can enable the project leader to 'fine tune' the support being given to the end user. However, even with sophisticated authoring systems and computer-managed learning, there is no substitute for a thorough assessment of the end user's capabilities and a well-specified training needs analysis; these activities will be the cornerstone of successful training.

11.5 Implementation relationships

Apart from the project team, there are two broad classes of people involved in implementation:

● users
● influencers

At first glance, it may seem to the project leader and the implementation team that the only really important relationship is between the team and the end user. This statement needs closer examination to decide who is a user. A user may be a person actually operating the sytem, generating input and extracting information. However, these users do not operate in isolation from the rest of the organization, and others may require help from them in the form of summaries, reports or analyses – we may call them 'indirect users'. The use of sales invoice and payment information, for example, from the finance department may alert the sales and marketing functions to cash-flow difficulties even though the volume of sales may appear high. Sales and marketing departments may wish to see information at particular times or when particular critical limits are reached. They may therefore make useful contributions to the introduction and smooth running of the system.

The implementation team also have to build up a good relationship with those who may influence the successful outcome of the implementation. The case of top management is covered fully later, but we should remember the role of staff representatives in every type of organization who are concerened about the effects of information technology. Their willingness to allow their members to participate in pilot trials and prototype systems can affect the performance of the implementation.

People are also influenced by the professional bodies to which they belong, and what they support and recommend will influence their members' attitudes within the firm to equipment, software standards and so on.

Personnel and training departments sit, depending on the organization, between the influencers and the implementers. According to their inclination, they may be closely involved in implementation because of their skills in training needs analysis and course preparation. Therefore, the project leader should have already enlisted their help and support, to prepare them for an active role, long before implementation starts.

We have already indicated that there are several parties involved in implementation. While the system must itself be technically sound, managing the implementation is really about managing relationships. These relationships are between:

● people
● people and the technology
● people, their work and the organization

11.5.1 People

We have already seen that it is crucial for the project leader to secure the support of top management for the implementation. First of all, they will ensure, by their sometimes uncomfortable questions, that the system will provide the business benefits for their organization. Secondly, they will have ultimate control over resources, especially financial ones. It is they who can decide whether the project team can buy that latest release of equipment and software that was not planned in the original budget. However, they cannot be expected to support such requests if their own understanding has not grown in line with the development of the project or if no effort has been made to keep them not only informed but interested and excited about progress. Finally, top management's attitude will determine the attitude towards the system throughout the organization. If they give visible support to the introduction of the system because they are enthusiastic and see the benefits, then their managers will also give it a high priority in relation to their own normal daily workload, and this will encourage the rest of the staff to do likewise.

Staff representatives may also be very positive or negative forces for change. They will probably be a focus for their members' concerns about the new technology and will want to ensure that they know the facts about the new system before committing the support of their membership to any plans. Therefore, the project leader must enter into a dialogue with them because, at the very least, their members will be needed to participate in pilot trials and prototype systems. Moreover, there will probably be positive contributions about the effects of the system on pay, grading, careers, shift rotas, maintenance and so on because these aspects will be of more immediate relevance to them than to the project team.

In the past, there has often been a tendency to design systems and then present them to the staff representatives as a pilot trial, at which point it is usually too late to make other than minor alterations. By involving staff representatives in the design and conditions relating to its trial and its evaluation, issues can be more freely explored without the feeling of irrevocable commitment. As understanding grows, discussions of the systems' implications can be based on experience rather than on trying to cater for every contingency that might result. Finally, a word of caution: in general, systems project leaders have little training or experience in dealing with staff representatives or trade union leaders and therefore, the appropriate personnel advisers and line managers should always take the lead in discussions in these areas.

The project leader must also manage the contributions of the end user to the system design and implementation. This can take many forms. Apart from discussing requirements with users, their managers and staff in different levels and functions within the organization, the project leader can involve people in different ways: designing and evaluating screen layouts and report formats, writing and evaluating user manuals and obtained their secondment to work on particular parts of the project are all ways of generating commitment.

The project leader can also facilitate the system's introduction by ensuring that the end users also have a feedback channel to those who are influential in its success. Top management may wish to know whether those who will eventually use the system are happy with it; this gives the users confidence that their voice will be heard. Trainers and personnel staff may also want this information to see if they need to put in extra resources, and the staff representatives may want 'objective' feedback direct from their membership.

11.5.2 People and technology

The interaction between the end user and the technology and the way the needs of this relationship are designed into the system is known under a variety of names such as human factors; ergonomics; man–machine interface. All these draw heavily on the knowledge and techniques of psychology, physiology, anthropometry (the measurement of the human body) and environmental measurement to ensure the final 'user friendliness' of the system. Many principles derived from these disciplines have already been incorporated into design guidelines and are almost taken for granted. However, it is still possible to see today badly designed products. Typical considerations for VDU operators include posture; the angle of the screen; character legibility; the amount of information that can be presented on a screen; brightness; heating; lighting and ventilation.

However, the user interface with the software is becoming more of an issue for further research. The structuring of a database in a way that matches the human thought process of classification and the manipulation of data through 'user friendly' query languages are both examples of this trend. The concept of 'fuzzy logic' is a new but still controversial extension of this to provide more realistic models of human reasoning for the computer to follow. Computers are designed to act on clear-cut instructions and responses: yes/no, on–off, binary decisions. Human beings, however, are not so rigid (some would say even simplistic) in their way of thinking, which allows for gradations of description ('very cold' or 'mildly warm') and for entities to belong to different categories (e.g. a home is both a building and a form of investment). 'Fuzzy logic' allows these forms of thinking to be built into computer programs.

Some other questions are also being raised:

- Can user-friendly systems allow the scanning and searching of databases in the same way that we can scan or skip through a book?
- What will designers do to enable machines to interface with the human voice in a way which is not constrained by the use of limited vocabularies for speech recognition?

It is the project leader's responsibility to see that up-to-date information of this kind is always available to the team to ensure they are aware of user-friendly developments when selecting equipment and software.

11.5.3 People and jobs

A fundamental part of implementation is to assess and monitor the effects of the system on people and the work they do. A simple task analysis by the organization and methods department is not adequate for this purpose. The effects of the system will be conditioned by people's perceptions, their attitudes and values about their work and their skills, just as much as by the job requirements stated in terms of duties and responsibilities in a conventional job description.

These psychological factors important in structuring a job and determining its content are many and varied. The principal ones to have emerged from applied psychological research over the years are those summarized by Clark in a report commissioned from Coopers & Lybrand Associates by the UK National Economic Development Office in 1983, and which are given below:

- variety of work
- level of decision-making
- skills employed
- degree of interest
- autonomy (control over workflow)
- supervision
- inter-personal relationships
- general working conditions

Moreover, the accepted boundaries between jobs may also be affected by the system. Jobs can interface with other jobs vertically or horizontally in the organizational hierarchy and within and between functions or sections in an organization. Furthermore, the job may also interface with other jobs outside the organization.

Obviously, the precise impact of technology on people's jobs will vary according to the individual organization and the type of system introduced. A way of handling this is outlined later; however, some generalizations can be made on the basis of accumulated experience. In clerical jobs, the tedium of lengthy and repetitive calculations, having to search out and collate information and performing continuous self-checking procedures to ensure accuracy can be removed from the individual and built into the system. When this can be done, the job-holder is often left with tasks requiring more discretion and initiative. These may sometimes be accompanied by a corresponding increase in salary and better career opportunities when these skills are demonstrated.

Secretarial work, aided by multi-function workstations, is also changing. Electronic mail and computer-held diaries may make the arrangement of meetings much easier, allowing secretaries to do more rewarding and challenging work. While this might previously have been the province of the manager, many tasks can now be delegated, including the manipulation of data and preparation of reports through the use of simple microcomputer-based application packages. The preparation of expense claims and administration of stationery are other tasks which can be made easier.

The supervisors' work may also change with the advent of a new system. Traditionally, supervisors have controlled large numbers of people, allocating and checking their daily work. As fewer people are needed for purely routine work, and individual prompts and checks are built into a new system, the supervisor achieves control by examining information provided by this system through transaction logs.

Therefore, as the role becomes less involved with the collation of control information, more time can be spent interpreting it and deciding on the most appropriate actions which need to be taken in the light of it. It may be used for identifying training needs or re-allocating workloads, for example.

The impact of systems on technical and professional specialists is less well documented and understood than on the types of staff just mentioned. While not wishing to generalize about computer-assisted design applications, the routine and time-consuming work involved in making design changes is reduced, leaving the professional more time to concentrate on the skilled aspects of the job such as decision-making or presenting a client with more design options. The development of systems analysis and design aids will have a similar impact on the project leader's activities. Furthermore, accountants and lawyers are turning increasingly to technology to stay competitive in their fees and provide a faster service to their clients.

Finally, managers too have been affected by technology in different ways. In some situations, they have gained work by avoiding the need to use intermediaries who may not always be able to respond immediately to their requests. One instance of this is the increasing use of spreadsheet-modelling packages or the learning of query languages to access databases or the use of keyword search procedures. All this means that they may become more intimately and immediately involved with their information instead of always needing someone to provide and interpret it for them. The latter situation may cause it to lose its effeciveness, since it cannot be provided on demand. At the same time, because a system is designed to contribute to the smooth running of a business, targets and performance of individual managers may become more visible and exposed to the scrutiny of their peers.

To sum up, the project leader must not only be aware of the effects of the system on different levels and types of jobs, but must make others aware of them too. Failure to do this will mean that the implementation of the new technology will appear to be continually springing surprises on the end users.

11.6 Organizational impact

The broad business benefits of the system for the organization will have been defined at the cost-justification stage. Some may be readily quantifiable and obvious to all: savings in staff; improved speed of customer service; reduction in customer complaints or queries; others can only be seen as long term, such as the ability to design more sophisticated products; and some will be very subjective, like a better corporate image.

While the impact of the system on individual jobs and the boundaries between them have already been covered, new technology may also affect the boundaries between the organization as a whole and its customers. For instance, where tour operators allow travel agents direct access to their own on-line enquiry system for the booking of holidays, work previously done by the tour operator's own staff can be transferred out of the organization. Conversely, with an integrated text- and data-processing system handling customers' invoices work previously done by an outside agency (e.g. to prepare paperwork for export goods) can be shifted back across the boundary between the two organizations and done 'in-house' on the computer. This is made possible because the computer can be programmed to provide guidelines and formulae for calculations, which need to be transferred on to shipping documentation.

11.6.1 Job change and job loss

Although the system will have been installed with the specific purpose of improving business performance, it will not be possible to predict exactly the effects on jobs and the precise form that this will take. The project leader should therefore set up a process to monitor users' experience with the system, providing them with guidelines to assess the impact it is having on their work. In that way, they can decide if they are sufficiently equipped for the job as it is now, using the new system, or whether they are likely to be in a role with diminished scope and responsibility. If it is a pre-requisite that the users are to contribute in this way, they must feel confident that their line management will support them and, if necessary, find them another job.

Questions which the project team should ask when assessing which jobs will be changed or lost are:

- Which tasks will be eliminated altogether by the system?
- Which tasks can be transferred to another job?
- Which jobs will remain but attract additional responsibilities?
- Will completely new tasks emerge which can be grouped as a new job?

Jobs which are likely to be lost are those where the input or generation of data can be done by the system itself or by another system to which it can be linked. Data-processing operators or clerk/typists may lose their jobs if a system with OCR is installed, for example. Similarly, machine operators may become redundant if numerically-controlled machines appear. Furthermore, jobs which are intermediary steps in relating one part of a process to another are also vulnerable if the new system compresses all the steps and allows, for example, all the transactions to be handled by a single person at a multi-function workstation.

11.6.2 Job creation

While the general trend is for systems to reduce the number of jobs involving routine or tedious work, new systems can create new types of job. An obvious example is where technical jobs related to the system appear within data processing and management services departments; these jobs involve the management and operation of the new technology. A similar, though less technically demanding, job may be created with 'super-secretaries' managing an admininstrative support function for professional staff, based on a multi-function workstation. Another category of new job relates to the interpretation and utilization of the information provided, such as identifying and contacting customers who may not be using the organization's system in order to generate more business. Finally, there are often new training posts to help staff and customers make use of the system. These jobs may become permanent, since after the system has settled down, new applications or updates to it are offered.

It is obviously motivating for staff if these new jobs can be filled internally, and it does have the advantage of conserving existing knowledge and company experience. However, certain skills related to the use of information technology may be in short supply and a fundamental task is to find the right persons to fill them. If the system is large, a skills audit may need to be done. There may not be anyone – or only very limited numbers – with the required skills. At this point, it may be necessary to use aptitude tests to spot those who may be trainable or successful in a different job. Experience suggests that it will probably be more satisfactory to use these aptitude tests with a mixture of psychological ones, which will indicate ability and interest, and internally designed interview procedures and schedules. All this needs to be done against a set of predetermined criteria, upon which the systems department and end user departments agree.

11.6.3 Information technology agreements

Clearly, information technology is an issue with which trades unions and staff associations will almost certainly want to be concerned. Naturally, the universal concerns are likely to be job security, pay and conditions, and health and safety, as well as training people in the use of the technology. Not surprisingly, the practice of producing guidelines for the introduction of new information technology varies according to the union and organization. Where firms already have agreements for technology in general, any procedures for dealing with information technology are typically just incorporated into them.

It is important to get these agreements into perspective. Unions are not concerned here with minor modifications but with systems which could have a significant impact on the way work is done. Employees should therefore be involved to identify what, from their viewpoint, may be possible problems. For this to take place, a company needs to agree to provide union representatives with information in a form which can be readily understood. In order to plan for the introduction of new technology, it is sometimes recommended that a standing committee (consisting of representatives of management and the unions) meets on a regular basis – say four times a year – to look at future developments and assess their likely implications.

In addition, a union may wish to nominate a technology representative (in the same way that it nominates a health and safety representative). The role may cover keeping up to date both with what is going on inside the company and outside it – for example, by discussions with other unions and by attending training courses; the representative may also be required to be part of a negotiating team as and when required. The advantage of this arrangement is that a good negotiator may not have the capacity nor inclination to understand the technicalities of a new system and, by the same token, a technically-minded person may not have good negotiating skills.

11.7 Guidelines for implementation

The importance of the role of top management to the system has been outlined already and reasons given for its criticality. Nevertheless, this support will not be fully effective unless it is visible to the rest of the organization. It may take many forms, depending on the nature and size of the system. A managing director may decide to sit on a steering committee if the proposed system is likely to fundamentally alter the way the whole organization conducts its business and if it is crucial to its position in the market place. Not all systems, of course, will warrant the managing director's or chief executive's attention in this way. Nonetheless, the project leader can ensure that continual reference is made to the proposed technology at company conferences, in house journals and internal training courses. These are all suitable ways for top management to express their commitment to the new system.

Throughout this chapter, continual reference has been made to the contributions users can make to the system, and their role in the analysis and design process was covered in *Practical Systems Design* (Pitman, 1984). Involving users will ensure that commitment to the technology is spread as widely as possible and that the system will be right. Nevertheless, obtaining their participation depends on their being properly motivated in the first place. This may not be an easy task, depending on the culture of the organization and its prior history of system introduction. Even so, there are ways of capitalizing on existing enthusiasm and generating more.

First, people want to belong to a winning team. Business objectives are set to ensure this can be readily understood at all levels. Maintaining a competitive edge and giving customers the best service are aspects of a successful organization with which people can easily identify. If the system can be seen to contribute to this, support will certainly be forthcoming. Secondly, if the system can be shown to make a user's job less tedious and at the same time offer scope for more discretion and challenge, it will increase motivation; as will the guarantee of no compulsory redundancy or the option of a transfer if people find they are not suited to their new job.

The project leader will need to ensure that the user's and systems staff work closely together and that their understanding of the new technology grows at the same pace. This will be of as much benefit to systems personnel as to users themselves so that they can contribute to the implementation process in a more positive way. A key individual to carry along with the team is the end user manager since staff will be encouraged and supported by this person. The project leader and the team must make sure that the manager is fully aware of and confident in his or her knowledge of the proposed system. As implementation may take place over an extended period, the team will need to seize every opportunity to maintain the manager's enthusiasm.

11.8 Summary

This chapter has stressed the importance of considering the implementation of technology in the wider context of organizational change. By doing this, it focused on the necessity of paying attention to aspects not directly related to the technical capability of the system but which exert a powerful influence on the success or otherwise of the implementation. Identifying and communicating business benefits to create motivation, being aware of likely pitfalls, being sensitive to relationships between people at all levels, the system and work are all vital ingredients for success.

Index